A Feathered River Across the Sky

By the same author

Of Prairie, Woods, and Waters: Two Centuries of Chicago Nature Writing
A Natural History of the Chicago Region
A Birder's Guide to the Chicago Region (with Lynne Carpenter)

Praise for *A Feathered River Across the Sky*

"The first major work in sixty years about the most famous extinct species since the dodo . . . equal parts natural history, elegy, and environmental outcry . . . Answering even basic questions about the passenger pigeon requires a sort of forensic ornithology, which gives *A Feathered River Across the Sky* an unexpected poignancy at the very points where it is most nature-nerdy."

—Jonathan Rosen, *New Yorker*

"Joel Greenberg, a Chicago-area naturalist and avid birder, has written a new account of the passenger pigeon's demise, *A Feathered River Across the Sky*. As Greenberg relates it, in calm, measured prose, it's a story of unremitting, wanton, continental-scale destruction."

—Elizabeth Kolbert, *New York Review of Books*

"[This] brilliant, important, haunting and poignant book . . . will forever change the way in which you think of pigeons (all birds, really) and about the natural world . . . The book describes, in vivid detail, forceful narrative and handsome illustrations, the history of this species and the factors that contributed to its extinction."

—Rick Kogan, *Chicago Tribune*

"There are a hundred ways to kill a species, and each extinction is both a tragedy and a drama deserving of its own story. This thoroughly researched and well-written account of the passenger pigeon is a definitive example of its genre."

—Edward O. Wilson, University Research Professor, Emeritus, Harvard University

"Throughout, Greenberg writes in a simple, straightforward way as if channeling E. B. White, leavening the depressing with the fascinating . . . [the book] doesn't just pack a punch—it packs a punch to the head, the heart, and the gut. Greenberg's *Feathered River across the Sky* is one of the most important books I've ever read."

—Laura Erickson, author of *101 Ways to Help Birds*

"An epic of life and death on a scale of billions . . . The world has never seen anything like the abundance and crash of the passenger pigeon. This astonishing boo[illegible] nd urgent story for today [illegible] nature writers, and his masterful command of this extraordinary topic makes this a must-read book for the ages." **—Kenn Kaufman, author of *Kingbird Highway***

"The human folly depicted here is as deep as the pigeons were numerous . . . Highly recommended." **—*Library Journal* (starred review)**

"Greenberg pulls together a wealth of material from myriad sources to describe the life and death of this species . . . [and] examines the larger lessons to be learned from such an ecological catastrophe—brought on by commercial exploitation and deforestations, among other causes—in this 'planet's sixth great episode of mass extinctions.' Greenberg has crafted a story that is both ennobling and fascinating." ***—Publishers Weekly***

"Combining genuine literary talent with a passion for research and synthesis, [Greenberg] has written a book that will henceforward be the first that I recommend to anyone wanting to learn more about these iconic birds. Greenberg provides broad-ranging information on all matters Passenger Pigeon, from prehistory to post-extinction pop culture . . . a rich and melancholy book." **—10000Birds.com**

"The first book about the [passenger pigeon] since A.W. Schorger's classic 1955 study *The Passenger Pigeon: Its History and Extinction*, and it is just as authoritative and just as important." ***—BirdWatching Magazine***

A Feathered River Across the Sky

The Passenger Pigeon's Flight to Extinction

JOEL GREENBERG

BLOOMSBURY
NEW YORK • LONDON • NEW DELHI • SYDNEY

To those who recorded what they saw; to those who collected the words; and to those whose love of beauty and life strive to make this story less likely to be repeated.

And to three others: Cindy Kerchmar, for enabling me to do this, and Renee and David Baade, whose support over the years has been extraordinary and deeply appreciated.

Published by Bloomsbury USA, New York
Bloomsbury is a trademark of Bloomsbury Publishing Plc

All papers used by Bloomsbury USA are natural, recyclable products made from wood grown in well-managed forests. The manufacturing processes conform to the environmental regulations of the country of origin.

LIBRARY OF CONGRESS CATALOGING-IN-PUBLICATION DATA

Greenberg, Joel (Joel R.)
A feathered river across the sky : the passenger pigeon's flight to extinction / Joel Greenberg.—First U.S. edition.
pages cm
Includes bibliographical references and index.
ISBN: 978-1-62040-534-5 (alk. paper)
1. Passenger pigeon. 2. Extinct birds. I. Title.
II. Title: Passenger pigeon's flight to extinction.
QL696.C6G74 2014
598.168—dc23
2013016927

First U.S. edition published 2014
This paperback edition published 2014

Paperback ISBN: 978-1-62040-536-9

1 3 5 7 9 10 8 6 4 2

Typeset by Westchester Book Group
Printed and bound in the U.S.A. by Thomson-Shore Inc., Dexter, Michigan

“Now, I’m sewing into the material
my red heart because the dead lately
have been a little noisy in my sleep . . .”

—Tom Crawford, “Prayer,”
The Names of Birds (2011)

Contents

Preface

My first recollection of reading about the extinct passenger pigeon was when I was in fourth grade at College Hill Elementary School in Skokie, Illinois. I checked out T. Gilbert Pearson's *Birds of America* and became so mesmerized I studied it relentlessly until the two-week period elapsed, after which I renewed it three consecutive times. Eventually, Mrs. Kelly, the librarian, encouraged me to sample other works that might interest me, including the Newbery Medal winner *Gay-Neck: The Story of a Pigeon*, about an individual *Columba livia* that flew messages on behalf of Allied troops in World War I. Although the story of the passenger pigeon crossed my consciousness while absorbing Edward Howe Forbush's account in Pearson, it became firmly lodged there in 1966 when I started birding as a twelve-year-old.

A year or two later my supportive parents responded to my request by giving me a hard-cover edition of A. W. Schorger's authoritative *The Passenger Pigeon: Its Natural History and Extinction*. Schorger was one of the country's premier historians of natural history. He spent forty years collecting thousands of sources related to passenger pigeons, starting his research at a period early enough when it was still possible to interview people who had known the living bird.

The pigeons not only affected the ecosystems of which they were undoubtedly a keystone species, but also the consciousness of the people who saw them. Accounts survive attesting to the presence of the pigeon hordes over every major city of Canada and the United States, from the Atlantic coast to the Missouri River. They were part of the cultural and economic development of these two nations. As passenger pigeon historian John French wrote, they were martyrs to our progress. So this really is a story of people as much as birds.

Indeed, the interaction between these two species is yet another element that makes the story of the passenger pigeon unlike any other. As late as 1860, one flight near Toronto likely exceeded one billion birds and maybe three billion. Forty years later the species was almost extinct, and by late afternoon on September 1, 1914, it was completely extinct when Martha, the last of her species, died in the Cincinnati Zoo. Human beings destroyed passenger pigeons almost every time they encountered them, and they used every imaginable device in the process. Unrelenting carnage reduced the population to the point where it began its inexorable spiral to obliteration. Whether a concerted effort could have reversed the decline and altered the outcome was a question asked far too late for any attempt to have even been tried.

I think that if an attempt had been mounted early enough to gather a sufficiently large and diverse group of breeding stock, they could almost certainly have survived, even if gone from the wild, because the bird bred readily in captivity. If a wild reproducing population had somehow survived a few more decades, it could have been protected by the strict conservation measures enacted in the 1930s—and based on scientific management, the species might still be with us, albeit in numbers much lower than billions. Modern Americans and Canadians coexist with cranes, waterfowl, and blackbirds that move across the landscape in flocks of many thousands or even millions, so why not passenger pigeons?

It is unusual when the exact date of an extinction is known with a strong degree of certainty. That the hundredth anniversary of Martha's death was fast approaching provided an impetus in my wanting to mark this event. It led to the writing of this book and a broader hope that this centenary could be a vehicle for informing the public about the bird and the importance that its story has to current conservation issues. In my research, I learned that others had the same idea, particularly ornithologist and pigeon scholar David Blockstein. We talked and began reaching out to other people and institutions. Among the first organizations I contacted were the Ohio Historical Society (which has on display the stuffed Buttons, one of the last known wild passenger pigeons), the Cincinnati Zoo (where Martha lived most of her life and died), and the Chicago Academy of Science's Peggy Notebaert Nature Museum. This eventually led to a series of conference calls involving an ever-expanding group of participants, and then Notebaert graciously hosted a meeting in February 2011 that drew about twenty organizations from across the eastern half of the country. Several others participated via conference calling. Project Passenger Pigeon emerged from that gathering, and as of September 2012, over 150 institutions were involved. Our goal is to use the centenary as a teaching moment to inform people about the passenger pigeon story and then to use that story as a portal into consideration of current issues related to extinction, sustainability, and the relationship between people and nature. It is hoped that this tragic extinction continues to engage people and to act as a cautionary tale so that it is not repeated.

CHAPTER I

Life of the Wanderer

> I have wandered through this land, just doing the best I can . . .
> And I can't help but wonder where I'm bound, where I'm bound,
> And I can't help but wonder where I'm bound.
>
> —TOM PAXTON, "I CAN'T HELP BUT WONDER WHERE I'M BOUND," 1964

Nothing in the human record suggests that there was ever another bird like the passenger pigeon. At the time that Europeans first arrived in North America, passenger pigeons likely numbered anywhere from three to five billion. It was the most abundant bird on the continent, if not the planet, and may well have comprised 25 to 40 percent of North America's bird life. When the flocks moved for migration or foraging, the earth below would be darkened by shadows for hours: famed naturalist John James Audubon recorded a pigeon flight along the Ohio River that eclipsed the sun for three days.

Only the firmament itself could absorb the legions of pigeons without being marked. When the birds descended to nest, feed, or roost, hundreds of millions of birds would sprawl across the landscape. The largest nesting on record took up 850 square miles. The birds congregated in numbers large enough to literally destroy the trees on which they gathered. Whether a site offered stately old oaks or scrubby willows, observers described the devastation as similar to that of tornadoes or hurricanes.

A passenger pigeon looked like a mourning dove on steroids: at fifteen to eighteen inches in length and ten to twelve ounces in weight, the pigeon

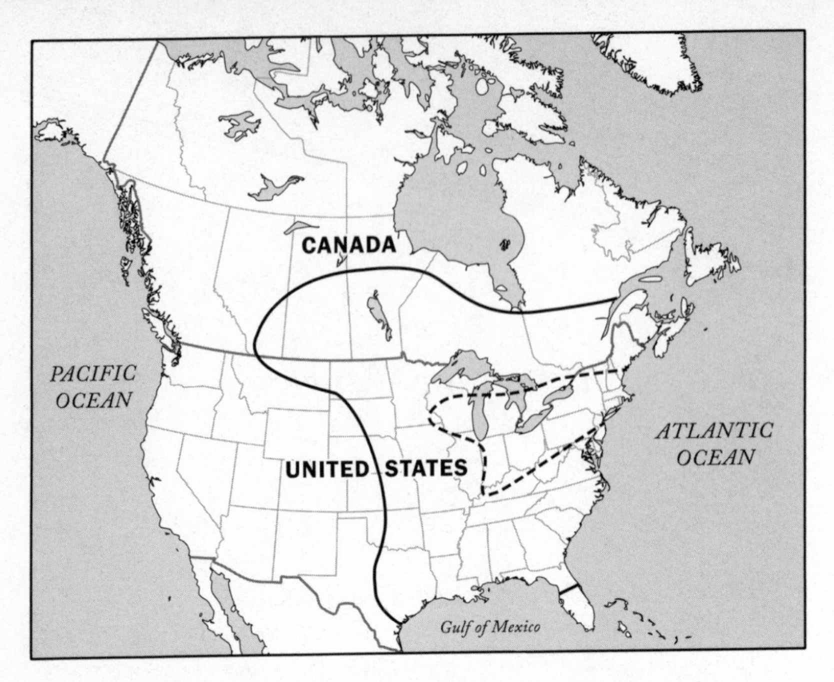

The range of the passenger pigeon in North America. Solid line encloses area of normal distribution while dotted line depicts area where the species usually nested in greatest abundance. © Gary Antonetti/Ortelius Design, based on a map in A. W. Schorger, *The Passenger Pigeon: Its Natural History and Extinction*

was one and a half times the size of the dove. (Of all the billions of passenger pigeons killed at the hand of modern humans, only a tiny number of fresh specimens were ever weighed, and the results recorded and published.) The male pigeon possessed slaty-blue and gray upper parts and a throat and breast of rich copper glazed with purple, while the female was a much drabber version throughout, with beige replacing copper. The Reverend E. C. Dixon witnessed one of the last large nestings in Wisconsin in 1882 and said it was a bird of "transcendent beauty. The male clad in a suit of gleaming iridescent, all but opalescent feathers . . . There was a sheen and brilliance far beyond mere simple color most birds display."

Wallace Craig, who studied members of a captive flock, wrote that the "voice and gesture" of the passenger pigeon was as different from that of other pigeons as it was in its manner of flocking and migration. He described the voice as "shrieks and chatters and clucks instead of cooing," although some observers did say that the species gave forth "coos." A very different impression of the bird's vocalization was conveyed by the New York minister who acknowledged that the pigeon had no song save for "a number of low notes, some of which are sounds that seem to be almost the soft breathing of the great trees."[1]

NOW HERE, NOW THERE: MIGRATION AND FORAGING FLIGHTS

Few North American birds could match the aerial mastery of the passenger pigeon. One of the few scientists who studied live passenger pigeons wrote of the species: "It was eminently a bird of flight," with a body that was "majestic, muscular, and trim." A life on the wing required specific adaptations to survive the grueling challenges. Connected to the deep keel of its breastbone, massive muscles powered the elongated wings, which beat the air like oars in the sea. Downward strokes kept the birds aloft while the slight turning on the upswing propelled the birds forward. Speed and agility marked this species in animation. John James Audubon said a lone passenger pigeon streaking through the forest "passes like a thought." He could never have guessed how prophetic that image would be.[2]

Passenger pigeons required speed and endurance, for they were nomads, ever searching for the forage that could best sustain their masses. Their movements were driven less by seasonal changes than this need for food. Although they generally headed north in spring and south in fall, for that reflects the continental pattern of greatest food availability, in some years members of the species lingered into December within sight of Hudson Bay. In April 1832 a nesting colony occupied thirty square miles of river bottom north of Columbus, Mississippi; and on January 1, 1876, a flock of many thousands of birds passed over Garrett County, Maryland.[3]

The editor of a hunting journal commented in 1913 on the wayfaring proclivities of the passenger pigeon: "The pigeon hosts were now here, now there; they were pilgrims and strangers, the gypsies of birdom." These pigeons were birds of passage, and their common name reflects this. But the point is made explicit in the redundancy of their scientific name. The genus *Ectopistes* is based on a Greek word for "wanderer," and the species *migratorius* is Latin for "one that migrates." This trait led the Narragansett of New England to call the pigeon *wushko'wh'an*, or the wanderer.[4]

Though striking as individuals, passenger pigeons most differed from other species in the size of their flocks and colonies. The flocks would take on every imaginable configuration. If a group of birds approached an obstruction, such as a hill, they might split in two and rejoin when the obstacle had been passed. Two oft-mentioned shapes were of long lines of birds, but in one they would have "narrow fronts" and in the other "broad fronts," up to forty birds deep. It was thought that the former was more common during shorter foraging flights, while the latter was more frequently adopted during migration. One possible explanation is that when the birds were covering extensive territory, a broad front better enabled them to assess the landscape for food, especially given their keen vision. When the birds were flying from established roosting or nesting sites to known locations of food, the slimmer flocks, generally comprising fewer birds, would be a quicker way of reaching the destination. But often the flocks would morph from one shape into another or into giant masses beyond precise categorization.[5]

When attacked by a hawk, the rivers of moving birds would drop to avoid the predator, but the notch in the line would continue long after the threat had ceased. Alexander Wilson surmised that after a while the inefficiencies of deep undulations in the thick cord of birds would dawn upon the members, who would "suddenly change their direction so that what was in column before, became an immense front, straightening all of its indentures." Perhaps there was no reason for the transfigurations at all, save for sheer caprice born of being part of a living multitude that dwarfed all else with which it shared the skies.[6]

The individual birds were packed into these immense flocks, seemingly with only enough space on either side to accommodate the pumping of their wings. Such rapidity of motion by each member makes the unbroken torsion of the throng just one more thing about this species that defies understanding: "The flight was very rapid and exceedingly graceful. While direction was held each bird appeared a law unto itself, swaying and twisting and veering, yet never colliding with one of its fellows." But as with most aspects of this bird, one can almost always find an exception. While traveling through Canada in the first few years of the nineteenth century, George Heriot reported that "when two columns, moving in opposite directions at the same height in the atmosphere, encounter each other, many of them fall to the ground, stunned by the rude shock communicated by this sudden collision."[7]

How fast these birds flew during migration can never be known with certainty, but based on a variety of evidence, some better than others, commentators from Alexander Wilson (1812) to A. W. Schorger (1955) accepted the speed of sixty miles per hour. Mourning doves have been clocked at forty-five to fifty-five miles an hour. Recently published genetic work reveals that pigeons in the genus *Patagioenas* are the closest living relatives to passenger pigeons. A member of that group, the band-tailed pigeon of western

North America, has been timed at 45 mph on leisurely flights, and a pre-migratory group exceeded the speed of a car traveling 70 mph.[8]

As the migrating pigeons reached their breeding or roosting destinations, or during foraging forays, they descended much closer to earth. One observer from Ontario tells of how the pigeons in their hurried flights just above the ground seemed to aim for him, but veered at the last moment: "It would almost seem sometimes as if they just tried to see how near they could come and get away successfully."[9]

One of the largest flights of passenger pigeons ever described in detail occurred at Fort Mississauga, Ontario, in May of what was probably 1860. Major W. Ross King was an English hunter and naturalist who spent three years traveling through Canada in pursuit of its "picturesque solitudes" and the wildlife that thrived within those prairies, forests, and waters so thinly settled by humans.[10] He had hoped to see one of those vast movements of passenger pigeons about which he had read so much, and he was not to be disappointed:

> Early in the morning I was apprised by my servant that an extraordinary flock of birds was passing over, such as he had never seen before. Hurrying out and ascending the grassy ramparts, I was perfectly amazed to behold the air filled, the sun obscured by millions of pigeons, not hovering about but darting onwards in a straight line with arrowy flight, in a vast mass a mile or more in breadth, and stretching before and behind as far as the eye could reach.
>
> Swiftly and steadily the column passed over with a rushing sound, and for hours continued in undiminished myriads advancing over the American forests in the eastern horizon, as the myriads that had passed were lost in the western sky.
>
> It was late in the afternoon before any decrease in the mass was perceptible, but they became gradually less dense as the day drew to a close . . . The duration of this flight being about fourteen hours, viz. from four a.m. to six p.m, the column (allowing a probable velocity of sixty miles an hour) could not have been less than three hundred miles in length, with an average breadth, as before stated of one mile.
>
> During the following day and for several days afterwards, they still continued flying over in immense though greatly diminished numbers, broken up into flocks and keeping much lower, possibly being weaker or younger birds.[11]

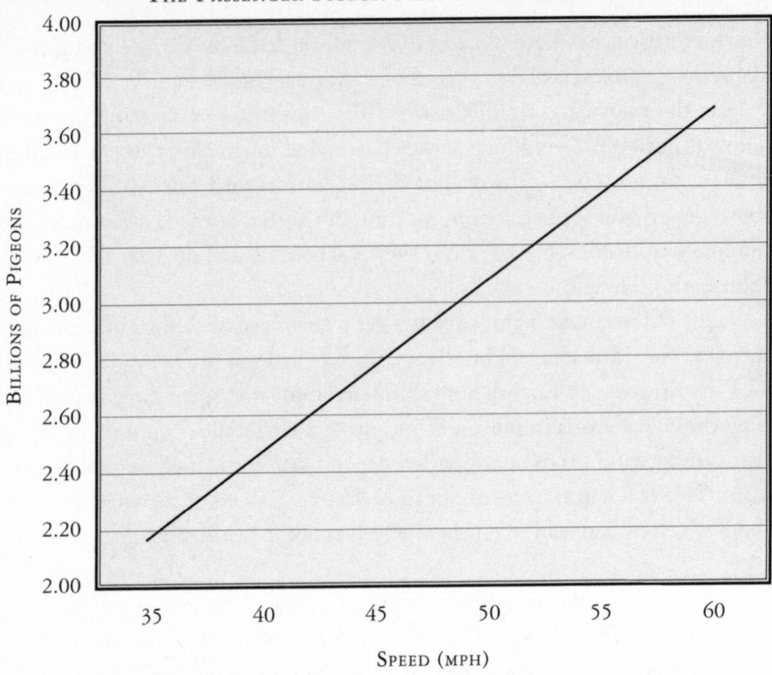

Graph showing the relationship between the size and speed of the passenger pigeon flight that passed over Fort Mississauga, Ontario (circa May 1860).
Based on graph created by Ken Brock

King never offered a numerical estimate, but Schorger, assigning two birds per square yard and a speed of sixty miles per hour, concludes that the flight involved an amazing 3,717,120,000 pigeons. At least three different scientists have each worked King's data in recent years and come up with the same results as Schorger, although doubting that the pigeons would be flying at 60 mph as a normal speed during migration. (Though as discussed above, band-tailed pigeons can at times fly considerably faster.) Ken Brock of Indiana University Northwest created a graph showing the numbers of birds at speeds from 35 to 60 mph. But even at 35 mph, closer to the speed at which mourning doves fly, which is more unlikely given that the passenger pigeon was a far more accomplished flier than the dove, King witnessed well over a billion birds passing over Fort Mississauga during the period of his observation.[12]

To make the size of King's flight easier to grasp, Schorger divided the size of the flocks into the acreage that produced it: nearly four billion birds drew upon the fecundity of 950 million acres. Four birds an acre can be grasped.

But it is misleading, too, because not every passenger pigeon was in that small section of Ontario. No one can know how many of these birds survived throughout the extensive range. To get a handle on that question would have required the employment of scientific techniques that neither existed nor would likely to have been supported given the lack of concern, appropriate resources, and institutions. Schorger's "guess" (his word) that three, and possibly up to five, billion passenger pigeons were present at the time Europeans first arrived on these shores seems to have been offered reluctantly, almost in the way of getting past a recurring question that can never be answered with any degree of confidence. (Schorger was meticulous and did not often engage in conjecture.) Whatever the number, this species enjoyed a population that may have exceeded that of every other bird on earth, and its aggregations surpassed in numbers those of every other terrestrial vertebrate on the continent.[13]

FUELING THE FORCE: WHAT PASSENGER PIGEONS ATE

Passenger pigeons fed on lots of things, including earthworms, snails, locusts, ants, and various insect larvae. But seeds and fruits made up a vast majority of their diet. When Pehr Kalm, the only student of the great Swedish taxonomist Carl Linnaeus to visit North America, traveled from New York to Canada in the summer of 1749, he "cut up some of the Pigeons which the French had shot" and discovered their crops to be filled with the fruit of the American elm. In 1924, a female passenger pigeon that had been killed many years before had a full crop that held twenty-five samaras of the sugar maple (*Acer saccharum*), all of which had the extensions detached at their base.[14]

A couple of incidents illustrate how eagerly passenger pigeons sought soft fruit. When a huge cohort of young pigeons left their nursery in a Mississippi swamp, they dispersed into nearby strawberry fields to feed. Local Indians explained their feelings on the matter: "You eat up our strawberries, we will eat you, Mr. Pigeon."[15]

Cherries drew the pigeons in this account by Henry Leonard. One summer day around 1870, two thousand people, many in bright costumes, gathered in Pigeon Cove, Massachusetts, to hear the Civil War general Benjamin Butler discourse on current politics. The speaker delivered his address from the shade of a large cherry tree, whose crown, adorned with ripe fruit, was visited daily by passenger pigeons: "In the usual afternoon feeding time of these birds a large flock of them alighted in every part of the tree; and although evidently surprised to find so great a company of men and women on the ground beneath them, and to hear the general's husky voice sending forth sentences like rattling shot, they made no haste to fly away."[16]

At least forty-two genera of wild plants were eaten, along with such domestic crops as buckwheat (their favorite), wheat, corn, rye, and hemp. Some of these became so thought of as pigeon food that their colloquial names reflected the connection. Margaret Mitchell searched the Canadian literature and compiled a list of seventeen such species including pigeon grass (*Setaria glauca*, *S. viridis*, and the nonnative *Verbena officinalis*), pigeonberry (*Rubus triflorus*, *Phytolacca decandra*, and *Aralia hispida)*, pigeon cherry (*Prunus pensylvanica*), pigeon plum (*Mitchella repens)*, pigeonweed (*Lithosperma repens*), and pigeon grape (*Vitis aestivalis*).[17]

When they had a choice, though, the birds overwhelmingly preferred hard mast, the forest nuts that vary in abundance from place to place and year to year. Although the birds would eat chestnuts, they were especially fond of beechnuts and acorns. (Even though chestnuts were a major component of many eastern forests, the historical record suggests that these trees played a small role in the life of passenger pigeons.) It is probable that a diet of these nutritious fruits during the time of breeding contributed significantly to reproductive success.

Each nut-producing tree has its own cycle. Beech, for example, produce heavy seed crops every two to eight years, making it one of the most variable of mast-producing trees in the forests of eastern North America. Despite this variability, large populations of beech seem to have masted every odd year throughout the 1870s at various places in Michigan, New York, and Pennsylvania. It was widely held that if the pigeons could choose a nut, they would opt for beechnuts, which are richer in nutrients than acorns.[18]

White oaks are also notoriously variable with good production every three to ten years, while black oaks are apt to mast at two- to three-year intervals. These cycles are based in genetics: large mast production insures that seed predators will be overwhelmed and some seeds will germinate; in low-mast years virtually all acorns are consumed. But they are also affected by both the amount of carbohydrates in the roots and the weather (a late frost could damage flowers and reduce the production of nuts in the fall).[19]

The acorns of some oaks, such as the white, mature in one year, dropping in the fall and likely sprouting by spring. Once sprouted, the nut retains little of its food value and ceases to be of much interest to the pigeons. Other oaks, such as the black and Hill's, need an extra year to develop and would thus remain suitable forage for a longer period including the critical spring nesting season. Roughly twenty species of oaks shared territory with passenger pigeons, and the birds likely fed on the acorns of all of them to a greater or lesser extent.[20]

"It is a wonder how pigeons can swallow acorns whole," wrote a perplexed Henry David Thoreau in his diary entry of September 13, 1859, "but they do." Millions of years of evolution led them to that capacity through the development of elastic and capacious mouths and throats. A mouth with a breadth of just over one quarter of an inch could be almost tripled due to a joint at each corner of the lower bill. After ingestion, up to a quarter of a pint of foodstuffs could be stored in the crop, a pocketlike area off the esophagus. At such times, the bulging neck almost doubled the size of the body. (When such a bird was shot, "it struck the ground with a rattle like a bag of marbles.") Eventually the material would be swallowed and then digested, taking up to twelve hours or more depending on what it was.[21]

A question that has survived beyond the pigeons themselves is how they knew where the mast was from year to year. Would the wintering flocks head toward the beech forests of Potter County, Pennsylvania, or the black oaks of Sparta, Wisconsin? Maybe this year it would be spring in the mixed woods of Benzie County, Michigan? It was thought by many in the nineteenth century that exploratory flocks of scouts would be sent out ahead to locate the choice areas and then return with the hordes. Simon Pokagon, a Pottawatomi chief and passenger pigeon historian (see chapter 3), broached the idea that passenger pigeons and other animals can communicate to each other over long distances. He suggested that something electrical in nature was involved.[22]

It is well-known that birds will suddenly appear at a newly present food source. In the tropics, the fruiting of a fig tree or the seeding of bamboo will draw an array of birds. A downpour in a parched landscape may well animate long-dormant invertebrates and even amphibians, which in turn quickly attract various birds. In North America, red crossbills are nomads that seek out the seeds of particular species of pine. Craig Benkman at the University of Wyoming has studied crossbills for years and believes that memory is a key piece of the puzzle in explaining how these kinds of birds operate. "I imagine that pigeon movements between areas of good mast were related to some knowledge of the distribution of the different mast trees and some recollection of current and past mast years," he explains. "If there was a poor crop in one region one year, they would more likely return to that area than, say, another area that just had a good mast year."[23]

The birds were also able to add to their knowledge. Crisscrossing almost a billion acres of range, the huge flocks of pigeons, possessed of superlative eyesight and often flying low, would be well equipped to assess the forage potential of vast areas. They would likely be able to remember at least to some extent what they had observed in the fall as a guide to their destination in the spring. Facilitating their search, mast areas tended to be extensive and so were easily located and able to accommodate lots of pigeons. This became a more serious problem for the birds as human activities whittled away at the forests, making it less certain that they would encounter areas of mast, at the same time that their populations were also decreasing. Yes, there would be fewer mouths to feed, but also fewer eyes to see. But each group of birds was not solely reliant on its own efforts: their knowledge of conditions would be augmented by the other flocks they met. "I suspect that a pigeon would be making a good decision to fly in the same direction that many other pigeons are moving in," says Benkman. "This would also explain the increase in and massive size of flocks." Vultures, wood pigeons, and great blue herons take cues from their neighbors in their selection of feeding sites. The ability to exchange information between individuals may be one of the reasons that some species form "large communal roosts."[24]

Passenger pigeons fed in a number of ways. While furiously beating their wings for balance, they would stretch their bodies from the branch on which they perched to pluck nuts within reach. This behavior led some early observers to surmise that the birds were dislodging the nuts with their wings, a behavior far too hazardous for any bird to undertake. A more spectacular sight ensued when large flocks descended to feed on acres of ground. As the mass of birds worked the area, there would be a constant leapfrogging, with the birds in the rear flying to the front to ensure they would receive their share. The birds moved in such a smooth and flowing manner they seemed to take on a fluidity that reminded some of water: "If they were approaching, there would be the appearance of a blue wave four or five feet high rolling toward you . . . When startled, their sudden flight would sound like rumbling thunder." Another likened the spectacle to "a rolling cylinder . . . its interior filled with flying leaves and grass." The scattering bits of vegetation made a "queer noise," not unlike the wind rustling dry foliage. It was a testament to the fecundity of these forests that despite the quantity of nuts consumed by the pigeons, the very ground they had methodically worked often yielded seedlings in the spring.[25]

A SCENE OF CONFUSION AND DESTRUCTION TOO STRANGE TO DESCRIBE: ROOSTING

To nest or to roost, passenger pigeons gathered in colonies of varying size. Accounts often make it difficult to know if the pigeon congregations were for nesting or roosting. If the goal was to nest, the birds needed woods and foraging territory that could sustain them between four and five weeks. Otherwise, the length of their stay was highly variable, depending on such things as the number of birds, the quantity of food that was available, and the intensity of human harassment. Some of these roosts would be used repeatedly, as for example one in Scott County, Indiana, that was visited for seventy-five years. More often than not they were situated in low, wet woods. The roosts might occupy as little as six acres or be as vast as 120 square miles or more. Free of the constraints imposed by nesting, the roosting birds could leave at first light and not return until nightfall or even later, often traveling in excess of a hundred miles a day. A report from Tennessee said that to procure the nearest beechnuts, the birds made a round-trip of four hundred miles a day.[26]

Pigeons jammed themselves into the larger roosts at densities that are difficult to imagine. Sixteen-year-old Sullivan Cook ventured into the Lodi Swamp of northern Ohio for his first pigeon hunt around 1845. He and his brother started their adventure early in the evening and began hearing the birds from three miles away. Pigeons began flushing as they entered the swamp, but they held their fire until they reached the main collection of birds: "As we approached this vast body of birds, which bent the alders flat to the ground, we could see every now and then ahead of us a small pyramid which looked like a haystack in the darkness, and as we approached what appeared to be this haystack, the frightened birds would fly from the bended alders . . . We now found these apparent haystacks were only small elms or willows completely loaded down with live birds."[27]

The largest roost known in Maryland was used from at least 1862 to 1872. Located in Allegany County, the site claimed six acres of alder swamp and was thought to draw birds from as far as fifty miles away. The place was packed: "So great was the number of birds that they were piled upon each other, in places, from one to two feet in depth."[28]

Perhaps even more impressive than the sight of the birds at roost was the late-afternoon flight when the pigeons returned en masse. C. W. Webber writes about an autumn gathering in southern Kentucky, close to where his father lived. He began his vigil an hour before sunset. A few small groups of pigeons appeared and disappeared, presaging a silence profound and eerie: "Everything seems to wait—listening for the great coming." A hawk and scattered crows, the latter flying without sound, headed toward the direction of the roost.[29]

The horses reacted first, pricking their ears at a newly discernible sound: "Is it a tornado coming? What a deep veiled roar! . . . The full burst of the deafening volume of that vast sound is borne upon you overwhelmingly with a current of fresh air strong enough to swerve you in the saddle. They are over us! We pause in speechless amazement. Half the sky is obscured . . . When will it cease? Is it one of the everlasting floods? We gaze until the real night is gathering around us."[30]

Texas was on the southern periphery of passenger pigeon range. In the fall of 1881, however, the state saw a massive incursion of these birds, probably from the big nesting that year in Pottawatomie County, Oklahoma. This was both the largest and the last of such flights to reach southern Texas. A study of these birds also provides a rare glimpse into the movements of one pigeon population as it shifted its roost from place to place.

By October 1881, the birds had formed a huge roost on the Colorado River near Austin. When the acorns ran out, they split into two groups, each going in a different direction along the river. The flock that headed upstream settled in shin-oak thickets near Burnet. Merciless persecution by hunters dogged the birds during their entire stay, but the abundance of acorns kept them in place from mid-November through December. Regarding the acorn supply, one commentator quipped that they were "as thick as hops on a vine or ticks on a neglected cow."[31]

The other flock roosted near Bastrop, just east of Austin. Here, too, they were killed by the thousands, but they stayed for only ten or twelve days before taking residence at Floresville in Wilson County, a bit to the south. The *San Antonio Daily News* of December 3, 1881, told this story of the roost with a straight face, although the second of the two alleged events is tough to swallow: "One citizen is reported to have killed 475 with a single stick, and another, while riding along with an overcoat on his back with large pockets, is said to have come out at the end of a lane with about three dozen of them in his pockets." By the end of the month, the birds were on the move again and joined their former flock mates, who by now had vacated Burnet. It is likely that these flocks and others came together in a gargantuan roost in Real County that remained intact until early February. This time they picked a relatively safe place, for the site was Frio Canyon, an area remote from human concentrations.[32]

Although differing from one another in many respects, these larger roosts shared at least one element. Their sheer volume imposed severe damage on the trees that supported them, sometimes making the timber valueless. Walter Rader, who grew up on a farm in Monroe County, Indiana, in the 1860s, recalled hearing the nightly crash of breaking limbs from a pigeon roost in a nearby maple grove.[33]

In one early Tennessee roost, where the "screaming noise" of the birds could be heard for six miles, they "settled on the high forest trees, which they cover in the same manner as bees in swarms cover a bush, being piled one on the other, from the lowest to topmost boughs, which so laden, are seen continually bending and falling with their crushing weight, and presenting a scene of confusion and destruction, too strange to describe, and too dangerous to be approached by either man or beast." Yet another old account (1801) from the Black River of the Mississippi Territory tells how the weight of the pigeons caused the branches to hang "down like the inverted bush of a broom." It describes how "a hickory tree, of more than a foot in diameter, was alighted on by so many of these birds, that its top was bent down to the ground, and its roots started a little on the opposite side, so as to raise a bank."[34]

Festooned over everything, pigeon dung distinguished the roosts as much as the dead and deformed trees. From that same roost on the Black River mentioned above, the author estimated that the amount of excrement carpeting the forest floor would fill "not only hundreds, but thousands of wagon loads." The chemicals released from this prodigious deposition of fecal matter eliminated the understory and killed the trees "as if they had been girdled." And the pungent aroma wafted for a mile, permeating the atmosphere with the scents reminiscent of a poultry farm.[35]

Western Kentucky hosted two major roosts of between four thousand and five thousand acres, plus several smaller ones. In the 1920s people could still provide firsthand information on the roosts back when they had been occupied by teeming pigeons. These elders could place the roosts with great exactitude: "a large piece of ground where a country store now stands called Longview," "around what is now the Jeff Davis Monument," and "on Bacon's Farm, one and a half miles South East of Ole Montgomery now." What struck them most about these former pigeon grounds was the fertility of the soil: "best land in Christian County" or "land is most productive . . . of any other land in Graves County." Eighty-eight-year-old Jasper Sisk summarized the fate of the old woods where the birds wintered: "They would roost in one place until they broke all the limbs off of the trees, then they would move to

Joining timber & treat it likewise, then fire would break out in the old Roost and Destroy the remainder of the timber. The land was then put into cultivation & considered the best farming land."[36]

WELCOME TO PIGEON CITY: NESTING

Generalizing about passenger pigeon nestings is difficult because they utilized virtually the entire diversity of wooded landscapes within their extensive range. They nested in single pairs, in small groups, in medium groups, and in vast colonies containing a hundred million birds or more. Pigeon cities tended to be longer than they were wide, which would have added considerable distance to potential food supplies over that required by a more compact configuration. Based on the dimensions of forty-seven nestings, Schorger came up with an average size of ten miles long and three miles wide.[37] The largest of all was the one that formed near Sparta, Wisconsin, in 1871. L-shaped, it spread across 850 square miles, although not every bit of that area contained nests. The 1878 Petoskey, Michigan, nesting encompassed over 200 square miles, and another in Huron County, Ontario, around 1870, was almost square at thirteen miles by eleven miles. And in 1823, a nesting of 180 square miles took place in upstate New York.

Although there are records of passenger pigeons nesting in most of the states and provinces they regularly inhabited, many of these involved nesting as single pairs and in small groups, about which little is known. A report from Portage la Prairie, Manitoba, described the local status of nesting pigeons: "They do not in this locality build in colonies, but place their nests singly, usually in small oaks." Unfortunately we will never know if those birds returned every year to form their little groups or whether in some subsequent year they joined the big nestings.[38]

Charles Douglas, a bird student and nurseryman from Waukegan, Illinois, reminisced to a friend about his experiences with the few passenger pigeons that used to nest nearby: "All of the nests I found were in the same place each year, and in groups of three or four, not many rods apart, in the big pines near Lake [Michigan] and in a small hard-wood grove a mile west." During the summer the birds entered his garden to eat cherries. Perhaps most noteworthy, the birds were allowed to nest and feed without disturbance.[39]

At the northern part of their range, the birds would often begin arriving in March. Some who knew the birds believed that they liked to be on territory just before the snow disappeared, exposing the bonanza of the previous fall's mast. Or they would scrounge bare patches of stream banks for worms. But just as many large nestings commenced when there was no snow at all.

According to some observers, small groups would arrive, then shortly thereafter be joined by the multitudes. But synchrony was an important element, so the birds tended to appear en masse. Unique in the passenger pigeon record is the amazing account of Chief Pokagon, who found himself engulfed by a huge flight of eager pigeons. In May of 1850, he was trapping on the upper reaches of the Manistee River in Michigan:

> One morning on leaving my wigwam I was startled by hearing a gurgling, rumbling sound, as though an army of horses laden with sleigh bells was advancing through the deep forests towards me. As I listened more intently I concluded that instead of the tramping of horses it was distant thunder; and yet the morning was clear, calm and beautiful. Nearer and nearer came the strange comingling sounds of sleigh bells, mixed with the rumbling of an approaching storm. While I gazed in wonder and astonishment, I beheld moving towards me in an unbroken front millions of pigeons, the first I had seen that season. They passed like a cloud through the branches of the big trees, through the underbrush and over the ground . . . Statue-like I stood, half-concealed by cedar boughs. They fluttered all about me, lighting on my head and shoulders; gently I caught two in my hands and carefully concealed them under my blanket.
>
> I now began to realize they were mating, preparatory to nesting. It was an event which I had long hoped to witness; so I sat down and carefully watched their movements, amid the greatest tumult. I tried to understand their strange language, and why they all chatted in concert . . . The trees were still filled with them sitting in pairs in convenient crotches of the limbs, now and then gently fluttering their half-spread wings and uttering to their mates those strange, bell-like wooing notes which I had mistaken for the ringing of bells in the distance.[40]

Almost all that is known about the actual mating of the passenger pigeon is based on Wallace Craig's observations of the caged birds kept at the University of Chicago from 1896 to 1907. Attempts were made to crossbreed several of the species in this pigeon collection, including passenger pigeons.

But passenger pigeons rarely mated with the other species as they acted rougher than ring-necked doves or rock pigeons. Either gender might initiate courtship. Sometimes the male passenger pigeon would land next to the female and press hard against her, often stretching his neck across hers in what the scientist called "hugging." (Perhaps a better term might be *necking*.) Craig notes, "Male presses over on perch against female, gives *keck* twice without raising wings, then preens inside wing, a sign of eros." She rebuffs his initial overtures, but eventually his subtle technique ignites her amorous spirit and she hugs him in return. Then, like all pigeons, they clasp bills, but in this species the contact is fleeting. She sits erect while he alights upon her back and clambers toward her neck, wings flapping all the while. Eventually she slumps, perhaps out of weariness, and lifts her wings to support him so that the act may be consummated. Upon separating, the male assumes "a position of fear," perhaps anticipating what is next on her agenda. Feathers out and neck withdrawn, he "clucks in a soft toneless voice." She replies in kind, before smacking him with her wings, another characteristic restricted to passenger pigeons. But she is just joshing, as their postures stiffen and they begin tickling each other for a few seconds.[41]

Some said passenger pigeons were monogamous, even to the point of forsaking all subsequent breeding should they lose their mate. This is unlikely to have been the case, but there would have been no way to know even if it were true. Nest building and egg laying generally took around three days. The nests were simple affairs consisting of seventy to a hundred twigs seemingly dashed together into the shape of a shallow cup to hold the egg. Males collected the branches from the colony floor and from farther afield, but should they drop their cargo, they would rarely retrieve it, although another bird might. Indeed, after a while, the ground looked as if it had been swept. Many observers were struck by the apparent flimsiness of the nests, one even calling them "shiftless." Some said that when standing directly underneath, it was possible to see the egg. Despite these poor impressions of the nests, however, the structures often persisted for years: the remains of some nests were still discernible up to twelve years after they were made.[42]

Passenger pigeons manifested little fussiness in where they placed their nests, using the local features and trees to their best advantage. Zebulon Pike found islands in the Mississippi River between what would be Illinois and Iowa where the birds nested on small trees in such abundance that "the most fervid imagination cannot conceive their numbers." In the Allegheny (New York spelling)/Allegany (Pennsylvania spelling) Mountains, the pigeons liked to nest on headwater streams, then spill downstream and up the wooded slopes as their numbers swelled. Several observers claimed the "nesting grounds were arranged with military precision." The borders, mostly straight lines, were clearly demarked, even if it meant that a tree on the edge would have nests on the inside face but be empty on the outside. This was a function, though, less of design than the birds' preference for a certain density. According to John French, arriving flocks would occupy their own wards, separated from others by "avenues . . . one mile or five miles wide" devoid of nests.[43]

Hardwoods were the trees of choice for these Pennsylvania nests, but open-growth hemlock also hosted their share. In other places nesting colonies tended to be in wet areas. Central Ohio's Bloody Run Swamp consisted of a low bog dominated by willow and poison sumac surrounding a higher middle where soft maples, swamp ashes, and white elms provided the nesting trees. The southernmost pigeon city on record formed along fifteen miles of the Tombigbee River in northern Mississippi. The mature forest consisted of such pigeon staples as oaks and beeches, as well as southern specialties such as sweet gums and cypresses. Different still was the great nesting north and east of Sparta, Wisconsin, in 1871. Most of these pigeons utilized a sandy area bristling with scrub oaks, but many birds spilled into more substantial timber of hemlocks and pines.

Breeding pigeons crammed their nests into almost every available space. John Josselyn wrote of his trips to New England in 1633 and 1663 and described traveling miles through pine forests loaded with passenger pigeon nests. A beech tree in a New Hampshire nesting colony in 1741 allegedly supported five hundred nests, while a visitor emerging from the Sparta site talked of seeing large trees with as many as four hundred nests. These were exceptional, if not exaggerations, but claims of sixty to one hundred were frequent.[44]

Dr. Gideon Lincecum is responsible for all that is known about the extraordinary Mississippi nesting. He and three companions rode horses into the swamp to check the progress after an eight-day absence: "Through all . . . the thirty square miles of that densely timbered bottom, from as high as one's head on horseback on the saplings, to the topmost limbs on the tallest trees, not a vacant spot where a nest could be crowded in, was to be found anywhere. We all searched with that object in view. The foliage of the trees had not yet unfolded but the packed and muffled up appearance of their tops made the swamp dark as midsummer."[45]

It might seem that with all the hunters running around nesting colonies, the number of eggs that the species laid would be a settled issue. But many otherwise thoroughly credible accounts are wrong on this point. Audubon and others said that two eggs were the norm, while Wilson and his faction claimed one. The best evidence makes it clear that only one egg was laid per nest: this is based on all but one (and that one is suspect) of the reports on captive birds and from a majority of the observers. Some of those who claimed two eggs, and many of these who also said that the birds nested multiple times during the season (a much harder issue to resolve in the absence of scientific techniques), were pigeon hunters who wanted to downplay concerns that the bird was becoming depleted. Honest error could also have crept in because in the crowded, chaotic pigeon cities one hen could have laid in another's nest.

Many others fudge the controversy by saying that the pigeon usually laid one egg, but on occasion would lay two. But in almost all instances, those claiming firsthand knowledge say it is either one or two. An exception is the writer who said that where the food supply was far from the nesting location, only one egg was laid. This might make sense except that the well-fed captive birds always laid one. J. P. Giraud, an early ornithologist, claimed that two eggs were laid but the oldest chick would toss the younger one out of the nest.[46]

My favorite account that attempts to shed light on this question cannot be reconciled even with itself. William Lehman visited the large 1870 nesting in McKean County, Pennsylvania, to obtain squabs (pigeon chicks) for his uncle. The uncle had created a spacious, well-watered, and vegetated park for the pigeons so that he might domesticate them. Lehman climbed small trees and in nearly every one claimed to have found two squabs per nest. He took fifty young birds, which he conveyed to his uncle. During their extended captivity, not one of these ever laid more than one egg per nesting.[47]

Typical of all pigeon species, the hens and toms divided up their domestic duties during the period of incubation. At dawn all the adult males would leave the colony to forage for several hours. They would return in midday to relieve the hens, who would then take leave for their own feeding forays. By late afternoon they would reappear, and the males would make one last flight for food before they came back at dusk. Although almost certainly incorrect, a common, but not universal, belief was that the birds would not feed within the nesting grounds so as to leave forage for the squabs. There is also evidence that the toms would often roost outside, though close to, the nesting areas. It was thought that they did this to minimize the chances that their great numbers would damage nests.

A passenger pigeon chick, or squab. Taken at Woods Hole, MA by J.G. Hubbard in August 1896, this is the only known photo of a living passenger pigeon squab. The bird was one of those in Professor Whitman's collection. Courtesy of Wisconsin Historical Society

The literature is replete with stunning depictions of the comings and goings of hens and toms. A morning exodus of males caught one group of Wisconsin hunters by surprise. The men had entered a portion of the huge 1871 nesting before light to "rake" the males as they embarked on their morning feeding round. With the first slight signs of dawn, small flocks of pigeons began to stir: "And then arose a roar, compared with which all previous noises ever heard are but lullabies, and which caused more than one of the expectant and excited party to drop their guns, and seek shelter behind and beneath the nearest trees. The sound was condensed terror. Imagine a thousand threshing machines running under full headway, with an equal quota of railroad trains passing through covered bridges—imagine these massed into a single flock, and you possibly have a faint conception of the terrific roar following the monstrous black cloud of pigeons as they passed in rapid flight in the gray light of morning, a few feet before our faces."[48]

An egg took two weeks to hatch, and brooding birds would squat with their bodies centered over the egg. When intruders appeared, the adult would tilt to one side to take a more cryptic stance and to free one wing should it be necessary to strike out at the threat. To shelter the squabs once they hatched, the adults would often use one wing to hold the baby next to the body.[49] A Detroit newspaper from 1880 describes a baby pigeon in these colorful, if unflattering, terms: "a featherless, hairy, misshaped, ugly looking little wretch of a bird, which soon develops the voice of a horse-fiddle in bad repair and the digestive capacity of half a dozen 14-year old boys."[50]

Squabs fed on pigeon milk, a curdy substance resembling loose rice pudding that originates in the crop of both the toms and hens. In pigeons the crop lining sloughs off to produce this highly nutritious milk that spurs rapid growth of the young. Analysis of this material shows that it is richer in protein and fat than human or cow's milk. Later, the milk is mixed with adult food. The adults opened their mouths wide, and the young stuck their bills into the corners to feed. Should the milk-laden squabs fall (or get pushed by the myriad of hunters who sought them out), they would often splat on impact like rotten fruit.[51]

It was widely reported that both hens and toms would feed orphans. To the extent that this behavior occurred, it was due either to confusion or because the accumulation of milk would be an irritant to the adult; at least one writer said that the failure to dispense with the substance could even prove fatal, although that seems incredible. That the orphaned squabs would be cared for was a favorite claim of the pigeon hunters, for it minimized the impact of the killings. It is indicative of their mind-set that Edward Martin, a game dealer dubbed Pigeon Butcher by author William Mershon, offered this story as evidence of his claim that adults would feed "squabs indiscriminately": "I may mention that one of the men in my employ this year . . . in one afternoon shot and killed six hen pigeons that came to find the one squab in the same nest." The field of public relations was evidently still in its infancy.[52]

The adults would feed the young for about two weeks. Then there came a day when the adults, finished with nesting, would rise in a dense cloud and evacuate the colony, leaving behind the roly-poly squabs to fend for themselves. They must have been quite a sight, for the youngsters "wander around like drunken men for three or four days." In the eyes of another, however, their movements were inexorable as they hiked their way to strength and flight: "If they came to a road they crossed it; a stream, they flew over; or they fell exhausted into the water and, flapping their wings, swam to the other shore and ran on until night." After another few days, they, too, could take to the air and form flocks that joined the adults. It is likely that this abandonment was due to human disturbance and that few if any of the younger chicks survived.[53]

Despite what would seem to be the overwhelming chaos of these mass nestings, the adults more often than not managed to return to their own nests. By what means they accomplished this amazing feat is not known. Yet another of the great questions that can never be answered regarding the life history of this species is how many times a year they bred. The literature provides a range of answers that is literally as great as it can possibly be: from once to monthly. One author said eleven, giving the birds a month off for travel. Without marking birds, no one could know for sure. Schorger concluded that they nested only once a year. This is based on his examination of the record and failure to find large nestings in the same year that were far enough apart in time for the same birds to be involved. He also points to the prodigious quantities of mast that would have had to be available in the same place to support two nestings in the same year. He does acknowledge one possible exception, and that was the Petoskey, Michigan, nesting of 1878, where birds did linger long enough to have nested twice. There are also records of single nests being found in August and September, but these may well have been second attempts after previous ones had failed or might have meant hatching-year birds were breeding for the first time. Some passenger pigeons might on occasion have had multiple broods in the same year, but it's unlikely that many of them did so with any regularity.[54]

MEMBERS OF AN ECOSYSTEM: RECONSTRUCTING THE PAST

> Thus it will be seen how the all-wise Creator, even in the case of these birds, has so wisely adapted the size of the food supply to the number of mouths to be fed.
>
> —PEHR KALM

Billions of passenger pigeons moving across the Canadian and American landscapes were subject to the constraints imposed on all species. A number of known factors took their toll on birds, although the principal limitation was the forage base itself. As many birds as there were, the population could not exceed the capacity of their environment to sustain them. That environment, of course, held physical and biological forces that culled varying numbers of individuals.

Adverse weather has always been the bane of migratory birds. (One of the worst kills on record occurred during the Memorial Day weekend of 1976, when a storm claimed over two hundred thousand warblers, thrushes, and blue jays that were later discovered covering a short section of Lake Huron shoreline.) Although as well suited as any other species to withstand the vagaries of atmospheric forces, passenger pigeons were not immune. In the spring of 1740, several sea captains related a calamity the likes of which they had never before seen. They had just reached Philadelphia, having traversed three miles of floating passenger pigeons. The prevailing speculation claimed that inclement weather had forced the birds offshore and prevented their return. Then the onset of night and weariness drove them to the water, where they perished. Less than a century later, Henry Schoolcraft found the beaches of Lake Michigan strewn with "great numbers of the skeletons and half-consumed bodies of the pigeon, which, in crossing the lake, is often overtaken by severe tempests, and compelled to alight upon the water and drowned in entire flocks." Fog rather than wind proved most dangerous to passenger pigeons, particularly those experiencing their first migration, for the adults had the savvy to rise above the poor visibility into clear sky.[55]

Far from water, hardships could also await pigeons. An early April storm hit Sullivan County, Pennsylvania, in 1868 and dumped a foot of snow on the same day that the pigeons arrived. A large tract of beech trees nearby was often used as a nesting ground during springs following a mast year, as this was. Ordinarily, the snow would disappear in a day or two, but when temperatures stayed low for two weeks, pigeon forage remained buried. A few birds sought sustenance in sumac seeds, but most were unable to find any food at all. Birds too weak to fly and then corpses by the many thousands soon covered the entire area.[56]

Another cause of mortality was brought about by the pigeons themselves. The huge masses of pigeons provided protection from predators and helped locate food sources. However, the sheer volume of their neighbors resulted in the death of large numbers of birds. Just think of the roosting and nesting colonies brimming with pigeons. Then imagine the destruction that would ensue when tree limbs, or at times entire trees, snapped and plummeted to the ground, crushing hundreds if not thousands of birds. When flocks descended to drink, at times the birds that landed first would drown under the weight of newcomers.

Predators took their share as well. Animals devoid of wings or the social structure and technology of *Homo sapiens* could not follow the big flocks, but other predatory mammals in the vicinity of a gathering helped themselves to the temporary buffet. Martins, mink, weasels, and raccoons would climb trees to procure eggs, squabs, and adults. Although I have not seen gray foxes specifically mentioned, they routinely scamper up trees and might well have preyed on pigeons. Black bears, bobcats, and mountain lions, all proficient climbers, joined the landlubbers, such as wolves and skunks, in seeking the dead or flightless birds that accumulated on the ground.[57]

Barred and great horned owls (and perhaps other species of owls as well) took pigeons at night. During the day, vultures would eat dead and wounded adults, as well as the vulnerable squabs displaced from their nests. The principal diurnal predators, however, were the two groups of bird-eating hawks. Five species of hawks native to eastern North America rely almost exclusively on birds, and each of them would go after pigeons if given the chance. But the fortunes of the three largest, the northern goshawk, Cooper's hawk, and peregrine falcon, were likely tied to that of the pigeons. The northern goshawk and Cooper's hawk (as well as the smaller sharp-shinned hawk) are accipiters, with short, broad wings and long tails that endow them with superior maneuverability. They relentlessly pursue their targets, following every dodge and cut as their prey desperately try to escape. Should the victim seek the shelter of a tangled brush pile, the hawk is apt to follow. Goshawks are the most brazen and unrelenting of the continent's avian predators. One story has it that as a chicken frantically tried to elude a pursuing goshawk, it sought security under the skirts of a startled woman, whose agitation no doubt soared when the goshawk followed close behind. On another occasion, a goshawk pursued a hen through an open doorway into a kitchen and caught the fowl even as the tenant and his daughter stared in amazement a few feet away. Audubon wrote of a goshawk as it made predatory forays above the trees and encountered a flock of pigeons: "He immediately gives chase, soon overtakes them, and forcing his way into the very center of the flock . . . then you see him emerging with a bird in his talons, and diving towards the depth of the forest to feed."[58]

Goshawks used to summer or breed as far south as northern Pennsylvania and southern Wisconsin. No data exist directly linking the events, but the pigeons disappeared and goshawks as nesting birds retreated much farther north. Today, they are but uncommon-to-rare fall or winter visitors to these former haunts. As for Cooper's hawks, there seems to be less specific information. In recognition of its predilection for the pigeons, southerners called the Cooper's "the great pigeon hawk." Ohio ornithologist Milton Trautman connected the concentration of pigeons with the former abundance of the hawk. Old market hunters told him that Cooper's hawks had declined in numbers between 1860 and 1921. The first thirty or forty years of that period encompass the decades that saw the great collapse in pigeon numbers.[59]

Should a group of passenger pigeons seek to avoid the skulking accipiters and take to the open sky, they risked contending with the equally formidable peregrine falcon. With long, pointed wings and a thin tail, these raptors coursed through the air like bullets: a peregrine can attain 180 miles per hour in a dive. Wherever this cosmopolitan species occurs, it preys extensively on some species of pigeon. Avian ecologist Stanley Temple has worked with peregrine falcons throughout his career and speculates that passenger pigeons figured prominently in their diet, particularly when the eastern forests were still intact. "Falcons would have found pigeons flying above the canopy to be among the easiest of prey," he told me. "They were the perfect size and of the perfect behavior." Even a peregrine with closed eyes dashing through a compact mass of pigeons at breakneck speed could hardly miss. (Of course the wrong kind of hit could result in grave harm to the hawk.) Once the forests were opened up, however, peregrines enjoyed easier access to a much greater variety of prey, and Dr. Temple thinks this prevented a steep decline in peregrines following the decimation of the pigeons. But he has no doubt that there was a strong connection between the species.[60]

The nineteenth-century ornithologist George Bird Grinnell tells of a peregrine falcon in hot pursuit of a passenger pigeon across the South Dakota plains. To elude its tormentor, the pigeon used Grinnell and his party as cover, even landing on the back of a horse. These maneuvers succeeded in confusing the falcon, for it lost sight of its quarry and had to ascend to reevaluate the scene. The pigeon, perched on the horse, could not see the falcon for a few moments. It then looked up, saw the raptor hovering, and made the mistake of launching forth in what proved a fruitless race against its faster adversary.[61]

Mast anchored the forested realm of the passenger pigeon, ecosystems providing sustenance for a range of creatures including filbert weevils (*Curculio*), blue jays, wild turkeys, and ninety species of mammals. Of the mammals, white-tailed deer are preeminent in how they impact the woods, and ecologists consider them to be a keystone species. So would, arguably, have been passenger pigeons: the two species would surely have affected each other, and the extinction of the bird no doubt altered the life of the deer. Other important mammals include mice, squirrels, and black bears. In poor mast years these species would decline or be forced to move on to other areas. Eastern gray squirrels in the 1800s were known to embark on great movements by the thousands. Black bears would also emigrate on a lesser scale. But when the nuts were thick, the local animals would prosper, although they had to compete with those vast flocks of pesky interlopers: passenger pigeons. This would later be the case with domestic swine, which in many parts of the country were allowed to forage in the forests, saving the expense of having to fatten them up on grain.[62]

It is impossible that billions of passenger pigeons did not affect the ecosystems that provided them with sustenance and quarters. Nor is it possible that their elimination in such a short time did not also impact those forests. Unfortunately, no one took much notice when the study of living passenger pigeons might have shed light on the many connections that had evolved over the history of the species. Only in hindsight have scientists attempted to explore the subject.

Imagine a show with two or three major characters and numerous ancillary roles. The principal actors would play off each other, as well as the lesser characters, who would in turn form relationships among themselves. Events trigger consequences that rattle around through the story, to be revealed at unexpected times or places. Now remove one of the stars and watch how events unfold. If you had never seen the original but only experienced the second version, you could only guess how that missing character would have affected the remaining players. In the case of the passenger pigeon, it is probably the best you can do. And if you subtract yet another member of the original cast (mast), the American chestnut, and add a major villain, the nonnative and highly destructive gypsy moth, the effort of reconstruction becomes even more difficult and speculative.[63]

The eastern forests started taking form when the icy cloak of the last glacial incursion melted about ten to fifteen thousand years ago. Vegetation began to reclaim the now bare ground. Trees pushed northward at rates and in patterns preserved by the records written in fossilized pollen. A seeming anomaly of that record is that the nut-bearing trees, with heavy, rounded fruit, expanded at the same pace as maples, elms, and pines, endowed with lighter fruit, some of which had wings that could ride the wind. (Another riddle involved the patchy distribution of beech, a species absent from large areas in between populations.) Two possible explanations were offered: either the hard-mast trees enjoyed earlier seed production and thus had more generations to compensate for shorter increments of expansion, or the seeds were moved by animals. Of course, these are not mutually exclusive.

Various calculations revealed, however, that the large seeds somehow moved a greater distance per generation than did the light seeds. This left animals as the likely source of the power that enabled these fruits to expand their range. The flaw here, though, is that unlike humans with their Johnny Appleseeds who deliberately relocate living plants, the continent's indigenous wildlife tended to collect seeds for food, thus exposing the material to the destructive forces of the gastrointestinal tract. It is true that species such as eastern gray squirrels and fox squirrels are known to cache nuts underground, and although most are consumed, anywhere from 25 to 40 percent escape later detection and often germinate. Ecologist Sara Webb concluded, however, that passenger pigeons provided a much more likely form of transport allowing the spread of mast trees. While the vast majority of the nuts consumed by pigeons would be digested, it might take up to twelve hours for such armored fruit to be broken down. During that time, some number of mobile pigeons would die. Given the billions of birds, and the millennia that transpired, enough seeds could germinate to create forests.[64]

Once these forests were established, passenger pigeons continued exerting an influence on their components and form, particularly as they promoted white oaks over red oaks. As stated earlier, the birds preferred the acorns of the red oak group to those of the white oaks. Heavy predation in this way favored the white oaks and probably contributed, along with other factors, to their becoming the dominant oaks in many parts of the eastern United States. Indeed, many forests of Ohio, Pennsylvania, New England, and New York consisted almost exclusively of white oaks. Conversely, the elimination of the pigeons may well have enabled red oaks to spread at the expense of the whites.[65]

The pigeons impacted the landscape in other ways, too, by imposing a regimen of disturbances through their nesting and roosting activities. We have already seen how the masses of birds would break branches and even topple trees, thereby opening up the canopy and creating tinder. Further, in areas used by the birds over time, excrement would blanket the ground like snow, chemically changing the soil through the deposition of nitrates and salt. Little, if any, understory survived, and the trees themselves became stressed. In another scenario, the canopy gaps would have aided the establishment of trees such as white oaks or red oaks that are inhibited by too much shade.[66]

These events would often act together in magnifying their total effects. If a tornado had recently visited a forest before its occupation by pigeons, damage by the birds might have been less than if the timing of the disturbances had been reversed, but still greater than had only one disaster arrived. But likely the most profound alterations brought about through pigeon-generated damage were an increase in the frequency and intensity of fires. Exposing the forest floor to light helped dry it out, and breaking branches created fuel to feed the flames. Fire plays a major role in what kind of vegetation thrives in an area. At the heavy end of the burning continuum, grasslands prevail, as grasses and many flowering plants have extensive root systems that can survive the immolation that is fatal to trees. On land that is less often burned, the thicker-barked oaks such as white and burr do well, although black oaks also prosper under such fire regimens. Where fires rarely reach, the red oaks, beech, and maples might predominate.[67]

Ecologist Reed Noss has suggested that passenger pigeon activity might have been important in the formation and maintenance of yet another ecosystem, the canebrakes, "one of the most enigmatic grasslands of the south." *Canebrakes* is the term used to describe the dense tracts of American bamboo or giant cane (*Arundinaria gigantea*), which can reach forty feet in height. They have decreased markedly to the point where they now rarely exist in large enough expanses to constitute a significant ecosystem. Their decline may well have led to the likely extinction of the Bachman's warbler and severe reductions in populations of Swainson's warblers.[68]

Canebrakes concentrated on rich alluvial soils and required a disturbance regime that kept them from being overwhelmed by trees. The soil quality proved to be excellent for agriculture, and Indians cut the cane to make room for their crops. But when the sites were abandoned, they would often be recolonized by the cane. This pattern of human disturbance, along with frequent flooding and anthropogenic fires, helped maintain the cane. But Noss points out that even before humans arrived in the American south there was enough cane to support red pandas, whose occurrence is documented in the fossil record. Therefore, other factors must have been present to promote the growth of the cane. Passenger pigeons, by opening the canopy, enriching the soil with their droppings, and increasing the frequency of fires, might well have been one such factor. In support of this idea, when white immigration ended native agriculture, reduced fires, lessened flooding, and destroyed passenger pigeons, canebrakes retreated as well.[69]

Beyond their impact on the forests and their relationships with certain hawks, the pigeons have been linked with several other species. Although intriguing and well conceived, these connections are based mostly on conjecture. One such connection involves the American burying beetle (*Nicrophorus americanus*). This once-widespread scavenger was reduced to a range consisting of two islands off Rhode Island (one a colony deliberately introduced for conservation purposes) and a few counties in eastern Oklahoma. The populations in between disappeared over fifty years beginning in the 1930s. This collapse has been called "one of the most disastrous declines of an insect's range ever to be recorded." In searching for explanations behind this, some scientists have looked at the ideal-size food for this one-inch beetle and wondered which among them have also suffered precipitous decreases. Two candidates have emerged: greater prairie chickens and passenger pigeons. While the beetle maintained its range long after the pigeon was gone, the loss of the two birds might have forced the beetle to rely on other carrion sources that over time also declined for varying reasons.[70]

A special kind of relationship that impacts people is that between mast, mice, deer, black-legged ticks (*Ixodes scapularis*), and the bacterium *Borrelia burgdorferi*, which causes Lyme disease. A bane to those who spend time out of doors, Lyme disease first manifests itself through fever, headache, fatigue, and a rash, but in advanced stages can affect the heart, joints, and central nervous system. I used to feel smug that being a nature nerd I was not likely to contract the virulent new diseases that began emerging in the 1970s. But then came Lyme and West Nile (carried by birds and spread to people primarily by *Culex* mosquitoes), and it is no longer even safe hanging out at tamarack bogs at dawn.

When mast is heavy during the autumn, white-footed mouse (*Peromyscus leucopus*) populations soar the following summer. Mice are the principal carriers of the organism that causes Lyme disease, so when mice are abundant, the ticks have increased opportunities to take a slug of mouse blood loaded with the bacterium. Understanding this fully requires looking at the four-stage life cycle of the black-legged tick. First, there is the egg, which just lies there in the leaf litter after being laid in spring or early summer. The larvae emerge in summer, usually free of the disease. After one blood meal from a host they molt into the nymph, which is when the tick is most likely to become infected. The odds that will happen increase if the host is a white-footed mouse, the premier reservoir of the disease. Ticks at the adult stage are the most finicky, strongly preferring white-tailed deer. Sometime in the fall, the adult will take one last sip from the deer it is riding, then drop off onto the ground to overwinter. The absence of mast prompts the deer to move around, thus spreading ticks to wherever it happens to be when the hangers-on decide to let go. Ornithologist David Blockstein has suggested that had passenger pigeons remained in large numbers, they would likely have reduced the forage available to mice, thereby keeping in check their populations and limiting the spread of Lyme disease. But nothing in nature is simple, and this idea has a flip side. Mice are also major predators of gypsy moths, so in outcompeting the mice for food, the pigeons would also have aided the spread of the moths. This would in turn have reduced oaks and other trees, thereby making fewer forests suitable to the pigeons.[71]

Passenger pigeons have also been linked to another human disease. Eastern encephalitis resides in birds and is spread to people via several species of mosquitoes, most particularly *Culiseta melanura.* This insect breeds in a specific microhabitat that is common in large swamps: small puddles that form under the trunks and roots of fallen trees. The twentieth century saw several major outbreaks in such states as Massachusetts (where over thirty people died) and New Jersey (twenty-two fatalities). Researchers at the Harvard School of Public Health suggested that the highly mobile passenger pigeons would have been less likely to become infected than such species as American robins, red-winged blackbirds, and common grackles, which have increased in the eastern swamps. Because they are more sedentary, these birds would have increased exposure to the infected mosquitoes.[72]

Birds are, of course, members of ecosystems, but to lice, birds themselves are ecosystems. Many lice are adapted to survive on a single species. If that bird dies out, so does the parasite. For a while it seemed that the passenger pigeon was host to two endemic lice that followed its host into extinction. In 1937, over two decades after Martha keeled over in her cage, R. O. Malcomson was examining feather lice in an old collection when he came upon three specimens collected from passenger pigeons. He concluded that they were unlike any other bird lice and named them *Columbicola extinctus*, or "extinct pigeon" louse. They were long and thin and likely inhabited the vanes of the wing feathers. Decades later J. Tendiero found what he described as a new species of feather louse that was also dependent on the passenger pigeon. He named it *Campanulotes defectus*.[73]

Lice will never engender much affection among the public, but these two species rode the passenger pigeon into some notoriety. Nigel Stork and H. C. Lyal used them as their principal examples of what they called "coextinctions" in a 1993 *Nature* article. They make the important point that every time a sexy species becomes extinct, the tiny, cryptic creatures that depended on it disappear as well. And sometimes the host doesn't have to disappear. When the last wild California condors and black-footed ferrets were taken into captivity, they were disinfected so as not to contaminate other individuals already in cages. Only later did scientists learn that among the parasites they destroyed were one from each of the more charismatic animals that occur on no others.

As is often the case with science (and one of the things that separates it from certain other epistemological systems), the truth can be inconvenient. In this instance, subsequent research determined that *Columbicola extinctus* was more catholic in its habitat than Malcomson thought: it also thrives on the very much alive band-tailed pigeon. The specimens he looked at had broken setae (hairlike bristles on some invertebrates); without that damage they were identical to the extant lice. As for *Campanulotes defectus*, it apparently never came close to passenger pigeons in life, for it was an Australian species that had somehow come to be mislabeled. So the story of the two parasites proves to be less than what it seemed, and in a way that is too bad. But the ledger of such things now shows two fewer extinctions, and the plight of puny parasites has a slightly higher profile.

CHAPTER 2

My Blood Shall Be Your Blood: Indians and Passenger Pigeons

> The Spirits of men came upon the earth seeking incarnation, among the birds and animals, with an appeal, "Ho, Elder Brother, the children have no bodies." But they were unheeded, until the pigeon came and answered: "Your children shall have bodies; my bones shall be their bones, my flesh their flesh, my blood their blood, and they shall see with my eyes."
>
> —SENECA LEGEND AS TOLD BY JOHN C. FRENCH

When, where, and how the first humans entered the western hemisphere are now matters of heated dispute. Most archaeologists would agree it was no later than about fifteen thousand years ago. These humans survived on the bounty of the land that awaited them, and they exploited the biological resources that best served their needs. The people who lived within range of the passenger pigeon, particularly in places where the birds nested regularly, incorporated the species into their diets and often into their social and religious lives as well. The story of the passenger pigeon's relationship with the early people of North America is found in the archaeological record written in bone.

Above all other animals, the white-tailed deer was a staple of the people who lived in the forests of the eastern United States and Canada. A single deer represented a large quantity of protein for the successful hunter and came with such bonuses as a valuable hide, bones, and often antlers. Among birds, the wild turkey was king. But when other food sources were to be easily had, humans sought variety. Much of this variety was available only seasonally. Prehistoric

Indians inhabiting western Michigan, for example, focused almost exclusively on lake sturgeon during the spawning period. Many of these traditions survived into historical times, ceasing only when white activities either prevented further access to the resource or so lowered prey populations there was nothing left to take.[1]

Among birds relied on for food by Native American groups within passenger pigeon range, only the wild turkey was of greater importance than the pigeon. Although passenger pigeon remains are frequent at some sites, they are lacking or in low numbers at others. In understanding this, it is important to note that the archaeological record reflects use by the people living in a given place, and not necessarily the abundance of the animal. With passenger pigeons, abundance was at best seasonal, and in some years the birds may have been present in small numbers or even absent.[2]

Beyond that, there is the matter of taphonomy, or what happens to a carcass. In northern Minnesota, for instance, beavers appear so often in prehistoric sites one might conclude that they were more important to those people than bison were to the Plains Indians. But the reality is that beaver bones are dense and preserve readily in the soils of that region. Bird bones, in contrast, are hollow and often don't preserve well. The remains of geese, turkeys, and ducks do show up in large numbers in some places, but they have larger bones than passenger pigeons and thus were more likely to be preserved.[3] Where Indians could exploit nesting passenger pigeons, they overwhelmingly favored the squabs, whose bones had not yet even hardened. Many squabs were cooked in soups and stews, which would have further reduced the resiliency of their bones. Finally, many of the young birds were eaten at the temporary camps set up at the nesting sites and never made it back to the permanent dwelling places.

However, at some archaeological excavations the pigeons do make up a significant portion of the faunal remains. Modoc Rock Shelter, a riverine site four thousand to six thousand years old in southwestern Illinois, proved to be a trove of animal remains. Over nine hundred bird bones representing at least fifty-six species were recovered, most of which were waterfowl. Of the remainder, sixty-seven bones were of turkey and forty-six of passenger pigeons. It is not an area where passenger pigeons likely nested, but noted zooarchaeologist Paul Parmalee surmised that hunters caught the pigeons as they migrated along the bluffs. The species may also have used the woods surrounding Modoc Rock Shelter as winter roosts.[4]

Meadowcroft Rockshelter, located in Washington County in western Pennsylvania, is one of the most extraordinary archaeological sites in the eastern half of the United States. Careful analysis of the evidence discovered there indicates that people used the site for at least sixteen thousand years and possibly as long as nineteen thousand years, making it perhaps the oldest known site of human habitation in all of North America. The shelter is under a cliff overhang sixty feet above Cross Creek and littered at the base with geologic rubble that has been accumulating for millennia. Hunters working the rich lands of the upper Ohio River valley used the place as a campsite where they could be protected from inclement weather. Remains have been recovered demonstrating the presence of human groups from Paleo-Indians through Mound Builders and those of the Woodland and Mississippian cultures that flourished until about the time of the European arrival.

Nonhuman predators, particularly birds of prey, also utilized the ledges and rocks to den, nest, and roost; their legacy is represented by the incredible deposits of small birds and mammals, which were not heavily exploited by their human competitors. This aggregation of biological relics assembled by such a wide variety of species over so many thousands of years led paleontologist John Guilday to reflect on the uniqueness of Meadowcroft: "It is a site archaeologists dream of. Here is a chance to look at a vanished fauna across the board; a chance to match culture and environment through the vagaries of time with the greatest of clarity."[5]

Archaeologists have identified 151 groups of animals among the remains, making Meadowcroft's archaeofauna one of the most diverse known from any North American site. Most of this total are from the pellets regurgitated by hawks and owls, with the latter contributing the large majority. Southern flying squirrels, passenger pigeons, and toads (species?) comprise two thirds of the vertebrate remains that have been identified. Of the 13,350 bird bones found, 75 percent are from the pigeons, representing at least 810 individuals. A high proportion of these are squabs, which rarely show up in human-generated remains. But the flightless chicks would have been perfect targets for the predatory birds that left behind most of these bones. If not for these hawks and owls, we would have no inkling that passenger pigeons nested in southwestern Pennsylvania, since they evidently stopped doing so before recorded history.[6]

Passenger pigeon bones are also scattered through various prehistoric sites farther east. Significant concentrations, for example, have been uncovered at Uren and Lawson, both in Ontario. A particularly rich discovery of passenger pigeon bones was made at Lamoka Lake in Schuyler County, New York, a site dating back forty-five hundred years, which was principally occupied as a hunting camp. Of the 359 bird bones found, 274 are from passenger pigeons. These represented forty-three individual pigeons. Two Pennsylvania sites, dating from 100 B.C. to A.D. 1190, yielded more wild turkey bones than any other bird remains, but in one, ruffed grouse were second and passenger pigeons third, while in the other the pigeon was second. Sheep Rock Shelter in Huntingdon County, Pennsylvania, is unusual, if not unique, for having been used by people for over nine thousand years and being situated near a pigeon nesting area. Pigeon bones have been found at all of the many layers of human occupation. At the oldest levels, remains of this species outnumbered those of all other birds. Archaeologists also discovered small quartz pebbles, which proved to be crop stones ingested by passenger pigeons and turkeys to aid digestion.[7]

Archaeologist H. Edwin Jackson identified 107 sites in the Southeast that had yielded at least one passenger pigeon bone. They span periods from the late Pleistocene to the early nineteenth century. The earliest sites of human habitation were from the Late Paleo-Indian (16,500–13,500 B.C.) stratum of Dust Cave in Alabama. But impressive quantities of passenger pigeons did not appear until the Late Woodland Period (A.D. 700–1000), for which a greater number of sites are known. The Oliver site in northwestern Tennessee produced 1,181 passenger pigeon bones, which constituted 87 percent of all the bird remains found there. Jackson says this is the first "clear evidence for the exploitation of roosting locations." That may be the case, but since mass nestings did occur during historic times in Kentucky, and at least once each in Oklahoma and Mississippi, the possibility exists that these were nesting sites.[8]

A paper in 1985 by Thomas Neumann fueled some debate by claiming that the paucity of passenger pigeon remains at prehistoric sites within its range demonstrates that the species existed in small numbers during prehistoric times due to predation and competition for food by Native Americans. Only after Indian populations plummeted due to the diseases brought by Europeans were passenger pigeons able to attain historical abundance. This increase in pigeons was further helped by the reduction of deer, turkey, and other nonhuman competitors due to the appetites of the new arrivals. The author, however, omitted many of the sites where pigeon remains have been found in large quantities. He also neglected to mention the many early European descriptions depicting vast flocks of passenger pigeons. And finally, he greatly overestimated the degree and impacts that human competition and predation would have had on the pigeon population. For these and other reasons, archaeologists have largely repudiated this assertion, but regrettably it gained traction by inclusion in Charles Mann's popular *1491*.[9]

Professor Jackson's work does suggest that pigeon numbers in the Southeast began to rise between A.D. 900 and 1000, about five hundred years before the appearance of Europeans. But the reasons for this increase are impossible to divine with certainty. Since nesting and roosting areas provided the best, if not only, opportunity to obtain large numbers of pigeons, perhaps people devoted more time to identify where these places were, figuring the extra effort would pay off in the easily obtained provender. Archaeologist Paul Parmalee offered the possibility that the paucity of passenger pigeon remains at Illinois sites might be "attributed to greater local abundance or availability of more preferred species."[10] A change in the availability of those other organisms might have spurred greater interest in passenger pigeons. Jackson believes it can be explained by a shift to agriculture by Native peoples that both reduced their consumption of mast and provided corn as a new food source for the pigeons. But the amount of corn that would have been available to the pigeons was dwarfed by the edibles provided by vast forests. In addition, the dominant nut eaten by woodland people was hickory, which was too big for pigeons. Acorns were not an important food source for humans in the eastern forests because they required a lot of preparation to remove the bitter tannin. And among the acorns, those from the white-oak types were sweeter and thus consumed more often by people, whereas passenger pigeons preferred the black-oak types that stayed on the trees for two years. Also, as we have seen, the pigeon diet was not restricted to mast.

Dr. Terry Martin of the Illinois State Museum has been a zooarchaeologist for over thirty years. His office is in a large warehouselike structure that houses both the museum's collection and most of its scientists. He has in a drawer a small box of passenger pigeon bones recovered from Fort Ouiatenon, an eighteenth-century fort in Tippecanoe County, Indiana. The French founded and ran the place, but a large Wea village stood nearby, ensuring frequent trade and visitation. (Wea were of the Miami people.) Among the contents of the box is a pigeon sternum, paper-thin but strong enough to anchor the powerful muscles that propelled the bird in its meanders through the sky. The flesh likely helped fuel people who coexisted with the pigeon for centuries if not millennia.

OF SUSTENANCE AND SPIRIT

> God sent a great storm and gathered all of these birds up and blew them all into the sea and drowned them all, and the reason for his doing this was that God had prepared this great number of food for the Indians,

and when he saw that the white man was starting to kill them, he did this thing. This is why the pigeons all disappeared at one time and were not killed out as the deer were.

—CHEROKEE TALE EXPLAINING THE EXTINCTION OF THE PASSENGER PIGEON, *INDIAN PIONEER HISTORY*, VOL. 32

To the Jesuits go the credit of being the first European note takers to penetrate the interior of what would become Ontario and Quebec, and the eastern and Midwestern parts of the United States. As their mission was to save souls, the priests spent time with the people they encountered and provided glimpses of their practices and beliefs. These accounts were recorded as letters or relations. In one dated 1636 and entitled "The Ideas of the Hurons Regarding the Nature and Condition of the Soul, Both in This Life and After Death," the author, Le Jeune, wrote, "At the feast of the Dead, which takes place about every twelve years, the souls quit the cemeteries, and in the opinion of some are changed into Turtledoves, which they pursue later in the woods, with bow and arrow, to broil and eat." He goes on to speak of the Huron notion that each being has two souls: "One separates itself from the body at death, yet remains in the cemetery until the feast of the Dead—after which it either changes into a turtledove, or, according to the most common belief, it goes away at once to the village of the souls."[11]

Two Indian tales, one from the Cherokee and the other from the Neutrals (a now-extinct Iroquois group who were concentrated in eastern Canada and upstate New York), are strikingly similar to each other and, indeed, to some parts of the Old Testament. It has been suggested that early contact with Europeans might have planted the idea of Noah's guiding dove. In the Cherokee language, the passenger pigeon was a *woyi.* A dearth of mast threatened birds and mammals with starvation throughout the mountains. But through its efforts, a *woyi* found one territory rich in nuts and thus rescued its neighbors. In the other story, the Neutral also faced starvation one winter. The culprit this time was the enraged manitou who locked all the waters securely in thick ice, thereby depriving people of the life-giving fish they depended on. But then a miraculous thing occurred. Clouds of pigeons arrived clutching branches of huckleberries in their beaks. As the birds passed over the village, they dropped the berries, leaving a trail from whence they had come and to where they were headed. Following the discarded fruit, the Indians trudged for days until they finally found the source of the berries and were saved.[12]

The Iroquois and the Cherokee also performed pigeon dances and songs.

The Iroquois pigeon dance could be enacted at any time of the year, but was the opening piece in the annual Maple Festival. The dance-song featured two leaders moving abreast and shaking horn rattles, while pairs of dancers, alternating between men and women, stomp-stepped in a big circle. As part of their Green Corn Festival, the Cherokee performed a dance that emulated the predator-prey relationship between the pigeon and the pigeon hawk (presumably merlin). Just as the falcon would dart into flocks to whisk away its victim, a dancer pretending to be the raptor would rush the line of his comrades and pick one off. Another dance was a pantomime depicting the hunting of pigeons at a roost.[13]

The Mi'Kmac of Nova Scotia saw the seasonal changes of stars illuminated in the northern sky as a chase in which Muin the bear sought to elude the seven kinds of birds that were hunting him. One of these bird hunters was Ples the passenger pigeon. Muin was the Big Dipper, a set of stars visible throughout the year, while Ples was part of the Boötes constellation, visible only during spring and summer. The group of seven become reduced in number as the stars they represent disappear for the season. Eventually the robin and the chickadee dispatch Muin, and the cycle begins again.[14]

Most of the European accounts focus on the ways that Indians hunted the pigeons. The Jesuits noted the use of nets in catching pigeons at salt springs in the Finger Lakes region, where up to eight hundred were taken at one haul. When Pehr Kalm arrived in 1749, about eighty years later, he observed Iroquois still using nets at the same salt springs: "The savages in Ondondago had built their huts on the sides of this salt field, and here they had erected sloping nests with a cord attachment leading to the huts where they were sitting; when the pigeons arrived in swarms to eat of this salty soil, the savages pulled the cords, inclosing them in the net, and thus at once secured the entire flock."[15]

At Green Bay, Wisconsin, in the mid-1600s, the local Indians strung their nets across long openings in the woods, cut specifically for this purpose. Low-flying pigeons entered the baglike nets, and a waiting hunter would yank the rope and cut off egress. On windy days when the birds streaked low through the forest, many hundreds could be caught in a few hours.[16]

It has been suggested that the Indian use of nets to catch passenger pigeons was learned from the Europeans. But such a notion seems unwarranted given that various of the nations employed this technology in catching fish long before they ever encountered white people. On the other hand, the intricate setups involving multiple live pigeons as bait that would be utilized by professional pigeon hunters in the 1800s was a Euro-American invention.

This distinction seems to be implied by a Cayuga in Ontario who recalled that his people never used "the white man's method of netting."[17]

One of the best accounts describing Native Americans and the pigeons comes from the writings of John Lawson as he explored the interior of the Carolinas in 1701. Unlike most European visitors, whose comments on the local avifauna were restricted to birds that were toothsome or conspicuous, Lawson compiled a much broader list. He distinguished 120 types, including two kinds of blackbirds, two kinds of buntings, thrushes, and swallows. Adding value to the list, he appended his own observations.[18]

He was tapped by the authorities to lead a small expedition of six Englishmen to reconnoiter a region whose people and features were almost unknown to the outside world. Accompanied by four Indians, the group left Charleston on December 28, 1700, in one large canoe. The party worked its way north and west, aiming for the Indian village of Sapona on the banks of the Yadkin River near the present-day city of Salisbury, North Carolina. Not too many days before they arrived, a number of their horses escaped, and while some members of the party went to fetch them, Lawson and others sought passenger pigeons, which were particularly numerous in that territory: "You may find several Indian Towns, of not above 17 Houses, that have more than 100 Gallons of Pigeons Oil, or Fat; they use it with Pulse, or Bread, as we do Butter, and making the Ground as white as a Sheet with their Dung. The Indians take a Light, and go among them in the Night, and bring away some thousands, killing them with long Poles, as they roost in the Trees. At this time of the Year, the Flocks, as they pass by, in great measure, obstruct the Light of the day."[19] Lawson makes clear that the pigeons do not nest in that area, so all that oil was extracted not from the pudgy squabs but leaner adults.

The Ho Chunks of Wisconsin (formerly known as the Winnebago, at least by whites) lived principally by hunting. Few animals were off-limits, but among those on the proscribed list were skunks, otters, gophers, horses, crows, ravens, and eagles. Passenger pigeons were "chief birds," members of the same group that included the Thunderbird, or "chief clan." Hunting for them occurred only when the chief decided to hold a feast, at which time many pigeons were served to the entire tribe. As in other tribes, the chief and his advisers likely waited for the right moment to hold the party: in that period just before the squabs could take wing, when they were at their most delectable. To procure the birds, Ho Chunks entered the rookery with long poles and pushed the squabs from the nest. The pigeons would then be cooked either directly over open flames or in a liquid. At other times, Ho Chunks would acquire birds by gathering those killed by storms.[20]

The arctic explorer and naturalist Sir John Richardson noted the importance of passenger pigeons to the Indians living on the shores of Lake Winnipeg when he visited Manitoba in 1827. They subsisted on lake sturgeon during the spring spawning runs, and then later in the season on ripened wild rice. But in that period when neither the fish or rice were available, the pigeons became the mainstay of their diets.[21]

A common thread that weaves through the relations that many tribes had toward passenger pigeons is the prohibition against taking adult birds at nestings. When the Potawatomie had access to a nesting, large numbers would establish temporary camps. The adult pigeons were largely spared, with most of the effort expended to procure squabs. These were either rendered to obtain oil for butter or preserved for future use by smoking and drying.

The Sioux also observed the proscription on shooting nesting adults, as Lafayette Bunnell discovered in 1842. Returning from a business trip, he let the current of the Mississippi River propel his boat homeward to La Crosse, Wisconsin. Not too far upstream of his destination, the eastern shore supported a thick growth of oak. As he approached, he "saw clouds of pigeons settling to roost, when crash, would fall an oak limb, and then a noise would follow like the letting off of steam." This second sound, he was soon to discover, was the chattering of countless birds. He was not the only observer, however, as he encountered several young Sioux intently studying the roost. They were trying to determine whether the birds had laid eggs. When Bunnell mentioned this later to a local friend, he was told "that the Indians never disturbed pigeons or ducks by shooting at them when nesting, and that the life of a man doing so would not be safe among the Sioux."[22]

Members of the Iroquois League were also prepared to enforce the ban on shooting nesting birds. Kalm related the story told by French hunters who had intended to get in some pigeon shooting but were discouraged by local Iroquois. The Indians began with friendly attempts at dissuasion, but these soon became threats when the hunters refused to change their plans. This restriction on the taking of the pigeons was bound up in religion and custom, but it served the practical purpose of keeping the nestings intact long enough for tribal members to fully exploit the resource. Frightening the birds off prematurely by allowing rampant gunfire in their midst would have reduced the take, to the potentially severe detriment of the people.

WE WILL BE FAT FROM DRINKING PIGEON OIL: HUNTS OF THE SENECA

> If I ever again hear that there is to be a pigeon hunt, I will try to go there. It is the best fun you ever saw. When we get back, people will not know us—we will be fat from eating squabs and drinking pigeon oil.
>
> —WILLIE GORDON, DESCENDANT OF SENECA CHIEF CORNPLANTER

Indian sites in New York dating from the late Archaic period (4,000 to 10,000 B.C.) have yielded passenger pigeon remains that have enabled archaeologists to reconstruct how the people organized themselves in utilizing the bird. When a pigeon nesting or roosting site was discovered, word spread and groups of Indians would converge on the area. Because these sites would change from year to year, people had no incentive to defend them against use by others, and this fostered societal cooperation. Further, the abundance of the birds also contributed to harmony between tribal clans since no one would come close to exhausting the squab supply. Sharing this information, therefore, with others had no downside and increased the likelihood that the same courtesy would be reciprocated in the future.[23] The Senecas, one of the original five tribes of the Iroquois, were living embodiments of that model.

The Seneca called the passenger pigeon *jah'gowa*, or "big bread." As a seasonal food, the flocks that returned in early spring provided relief from what otherwise was a lean time when community members were often forced to subsist on seed corn. The effort to procure the pigeons became an important activity between the harvesting of the maple syrup and the planting of crops. Except for a few elderly who lacked the stamina to make the trip, villages became empty as families embarked for the pigeon grounds. Much of the detail related to these annual hunts has been preserved through early contact with Europeans and the testimony of tribal members, indigenous and adopted. Seneca memory is further preserved through myths and ceremonies. The Seneca Maple Festival is still celebrated in late March, and the Pigeon Dance continues to be performed.

It was the custom among the Iroquois to replace their losses incurred in war by taking in captives, including settlers and soldiers. If they survived the initial brutality meted out to the vanquished, these newly acquired community members could rise quite high within the tribe. One such abductee was Horatio Jones, whose intelligence and physical strength made him a favorite of his tribe. He later returned to the white community and was appointed by George Washington as an agent and interpreter for the Iroquois, a position he retained for four decades. His talent in rendering Indian words into English earned him respect among the Seneca, and Chief Red Jacket, known for his eloquence, demanded that only Jones translate his orations.

Jones describes a pigeon nesting from the early 1780s. He and his adopted family were visiting his mother's brother, Chief Cornplanter, when a runner arrived with news: "Pigeons, pigeons!" The birds that year were nesting on the Genesee River.

> All was now bustle and confusion, and every person in the village who could bear the fatigue of travel at once set out for the Genesee . . . As the annual nesting of the pigeons was of great importance to the Indians . . . runners carried the news to every part of the Seneca country, and the inhabitants singly and in bands, came from as far east as Lake Seneca and as far north as Lake Ontario. Within a few days several hundred men, women, and children gathered in the locality of the pigeon woods.[24]

Things did not change much with respect to the Seneca and the pigeons over the next ninety years or so. What these hunts were like can be seen with great clarity due to the recollections of several participants who were interviewed by ethnologists. The group was split between those living on the land granted to Cornplanter, the last war chief of the Seneca, and those of the Allegheny Band. Willie Gordon, a trapper and noted bear hunter, said the pigeons arrived during late March and early April in flocks whose beating "wings sounded as thunder. They came as a plague of locusts and devoured every sprouting plant." When the birds appeared, tribal leaders dispatched scouts to locate the exact nesting grounds. The pigeons could be counted on to utilize forests within a tier of seven northern counties of western Pennsylvania. (The regularity of their presence in one general area runs counter to the periodicity of masting, but perhaps the forests were extensive and diverse enough to produce some quantity of nuts every year. And other types of forage may have been available.) The scouts closely monitored nestings and returned with newly hatched squabs. After inspecting the samples, the chiefs would proclaim "two weeks" or "three weeks." The condition of the chicks determined when the time was right to pull up stakes and head for pigeon city. Squabs just a few days shy of fledging were at their plumpest, and the fatty morsels were what drew the Seneca.[25]

Although it seems that almost everyone participated, no one organized the effort. People went "because the pigeon hunt was a good time—just like a fair or picnic." Lydia Bucktooth told of her family's decision to participate one year. Her mother had been ill and was reluctant to make the trip, but her father persisted, pointing out the vast nesting was worth seeing once before she died. Despite her infirmity, she acquiesced and joined the family as they canoed and then hiked over rugged terrain to their destination.

The pigeon aficionados traveled on almost every imaginable conveyance. Willie Gordon described the procession. Teams of oxen and horses pulled wagons filled with families and such assorted equipment as "axes, guns, cooking utensils . . . and barrels or bark casks for placing the squabs." Some boarded trains for part of the way, while others floated downstream in rafts or canoes. And lacking any other option, there were even those who made the whole trip on foot.

Even when tribes knew the condition of the hatchlings and could plan their trip to coincide with the optimal time, traveling still lacked strict organization. The families did select a leader, someone well respected and a good speaker who could deliver the customary prayers of thankfulness at the beginning and the end of the day. He made sure the procession to and from the pigeon grounds was orderly, and he divvied up camping sites among the families. (This person received no extra squabs or other compensation for his efforts.) When their destination was reached, they erected three-sided lean-tos to shelter them for the duration of the hunt. Hemlock boughs formed both the waterproof roof of the exterior and the soft beds tucked as deeply into the shanties as possible. A fire at the opening provided some warmth.

These practical necessities were embodied in myths that kept the rules intact for those who used them during the pigeon times. For us, they are a portal into the past. The Seneca assigned human attributes to many animals, and they saw an affinity with the passenger pigeons in that the birds also lived in large gatherings. Many tribes held that albino animals of all species possessed special powers, and the passenger pigeon was no exception. Chauncy Johnny John stated the rule regarding albino passenger pigeons: "Never disturb him, and never cut down a tree in which a white pigeon has nested."

In Seneca mythology, the White Pigeon, an aged albino, was chief of all the pigeons. In one version, he visited the venerated warrior Wild Cat, and in another, an upstanding young man who had not yet been sullied by the temptations of life. But in both stories the message was the same. To Wild Cat, White Pigeon said, "All the various tribes of birds had held council and decided that the wild pigeons would furnish a tribute to mankind, because their Maker had selected the wild pigeons for this important duty . . . Other birds had very little to give . . . because . . . [they lived] dispersed here and there, and could be obtained only with difficulty." Though the Maker had bestowed the pigeons as a gift unto mankind, the receipt of this largesse carried with it responsibilities: the pigeons could only be harvested at the proper time and in the proper way. Principally this meant that only the squabs could be killed; the adults on the nesting grounds would be spared so they could breed again and create new generations. And a "superintendent" was appointed over the community to ensure that the travels and hunts were conducted orderly and safely.

In the presence of the young man, White Pigeon repeated his assurances: "We have no protest to make against your coming to this place to obtain the young whose bodies resemble ours . . . You must know, too, that Our Maker has ordained that this our flesh shall be for the welfare and contentment of human beings dwelling on the earth." Although "full of joy" for being selected to serve humans in this way, White Pigeon explained that his people were displeased that so many of the humans present had forgotten Him who has created all and "think only evil things to please themselves." For their own good they should jettison their bad thoughts, for otherwise they will surely encounter misfortune. But White Pigeon brought other admonitions also: it was essential that the humans enjoy themselves, and the way to achieve that was through song and a particular dance; offerings of tobacco and personal artifacts must be made to the pigeons and the Creator; and "you and my people must unite . . . when we make this prayer and request of the Creator of our bodies."

While the Senecas were engaged with pigeons, all other business was put on hold. The Friends' Indian School in Quaker Bridge, New York, closed from mid-April to early May so the students could help their parents with the making of maple sugar and the collecting of squabs. In the spring of 1791, Colonel Thomas Proctor also learned how important pigeon time was to the Seneca. As a representative of the federal government, he was in the area to meet with the tribal chiefs to get help in contacting other groups farther south. But he was informed that there would be no convening that day for the pigeons had arrived and the Seneca considered it their duty to honor the Great Spirit for his generosity by holding a great celebration.

Proctor soon joined in the festivities, where he met a major chief for dinner. The women were dressed in their finery, adorned with silver, and they served a feast that relied heavily on pigeons:

> Some boiled, some stewed, and the mode of dishing them was, that a hank of six were tied with a deer's sinews around their necks, their bills pointing outwards; they were plucked but of pen feathers plenty remained; the inside was taken out, but it appeared from the soup made of them, that water had not touched them before. The repast being the best I had seen for a long time, I ate of it very heartily, and the entertainment was given with the appearance of much hospitality.

The hordes of both people and pigeons created the perfect venue for the top archers of each group to strut their stuff. They used bows made of white ash or hickory, seasoned with repeated applications of hot oil. Rawhide rolled into perfect roundness became the strings. The arrows had round heads for knocking the squabs from the nest.

Most of the squabs were gathered by a much less artful method than archery. After breakfast, families from both camps spread out through the nesting, in search of squabs. Each small unit worked independently of the others, and there was no sharing of the pigeons. But within families was a division of labor, as men cut the smaller trees and the women and children collected eggs from the now-accessible nests. As the squabs scurried in all directions, they were pursued by adults and children alike. The chicks would be dispatched by blows to the head, or if in the hand, by pinching their skulls or wrenching their necks. Processing consisted of removing the crops and intestines immediately, for otherwise the meat would quickly spoil. The squabs would then be salted in elm casks or stone crocks.

Most of the pigeons, though, were consumed at the camp and were generally boiled. To round out the meals, many would enjoy the delicacy of potatoes smeared with squab fat. The potatoes were purchased from the one white settler in the area who raised them next to his log cabin.

As rail expansion drove up the demand and prices for lumber, local landowners began objecting to the Seneca cutting down trees in the name of squab collection. Willie Gordon related an incident where mounted sheriff's police confronted his group, threatening to arrest them if they felled any more trees. But one of the elderly men, Jesse Logan, reminded the officers of the treaty that gave the Seneca the right to hunt, fish, take pigeons, and fell timber wherever they might be in twenty-two counties of New York and Pennsylvania. Another member of the group agreed to return to the town of Cornplanter to secure the treaty documents. He was on foot and so it took him two days to make the round-trip. Confronted with the evidence, the officers offered a compromise that the Seneca accepted: "If the Indians would leave the big trees so that the timber would be spared and just cut down the smaller ones, this would satisfy them."

A majority of the squabs not consumed on-site were sold to the dealers who were always in attendance. Lydia Bucktooth said her family joined the hunt mainly for the fun of it. As they traveled by boat and foot, lugging all of their camping and cooking gear was a huge chore; not being able to bring anything back anyway, they took advantage of the opportunity to make a little extra money by selling their birds to the traders. Alice White and her kin, on the other hand, traveled in a large oxcart capable of transporting barrels full of squabs. When they arrived back home, her generous father gave the birds to neighbors who had been unable to make the trip. Then there were those like Willie Gordon, who slogged home carrying ash-splint baskets hanging from straps across his back and shoulders and packed with dressed squabs.

William Fenton and Merle Deardorff, who so meticulously reconstructed these final hunts through the words of their informants, failed to specify the years that these events occurred. But human life spans being what they are, the happenings could not have been any earlier than the 1860s. The authors do say that these narratives are "the last shreds of Cornplanter Seneca ethnology." They are also the last depictions of what to the Seneca was "a regular event in the annual round of getting a living." This link between a people and a bird started before recorded time and would only conclude with the decimation of the latter. Living for one became more difficult; living for the other impossible.

CHAPTER 3

A Legacy of Awe

> March 26, 1780. At the Sunrise Services of Easter the brightness of the lovely morning was suddenly eclipsed by the passing overhead of countless multitudes of wild pigeons flying with their wanted swiftness from south to north.
>
> —BERNARD ADAM GRUBE, ENTRY IN CHURCH DIARY, LANCASTER COUNTY, PENNSYLVANIA

The two ships under the command of Jacques Cartier left Saint-Malo in Brittany bound for Newfoundland on April 20, 1534. Cartier hoped to find what inspired most explorers of the time: treasure and a Northwest Passage to Asia. He certainly failed in the latter and did not acquire any of the gold and jewels that were implied by the former. His most interesting cargo were the two young Iroquois he brought back with him and would return to their homeland on his second voyage. History has granted him laurels for being the first European to discover the St. Lawrence River, with help from his soon-to-be passengers, and to explore the islands and coastline of the Gulf of St. Lawrence.

On July 1, 1534, his party sailed south along the west coast of Prince Edward Island looking for a good harbor. None existed, but it was easy to access the shore anyway, as the water was not deep nor the shore steep: "We landed that day in four places to see the trees which are wonderfully beautiful and very fragrant . . . The soil where there are no trees is also very rich and is covered with peas, white and red gooseberry bushes, strawberries, raspberries, and wild oats like rye . . . There are many turtle-doves, wood-pigeons, and other birds."[1] Since only two kinds of pigeons inhabited eastern Canada, the "turtle-dove" is likely the mourning dove, and the "wood-pigeon" the passenger pigeon. This quote thus establishes Cartier as the first European to record the existence of the passenger pigeon.

A small sample of early accounts will give an inkling of what pigeon flights were like in the eastern part of the continent. Ralphe Hamor published a "true discourse" on Virginia in 1615 that tells of "*wilde Pigeons* (in winter beyond number or imagination, my selfe have seene three or four hours together flockes in the aire, so thicke that even they have shadowed the skie from us)." A Dutch chronicler writing about New York noted that passenger pigeons were the most common bird on Manhattan Island during the 1620s and, when massed in the air, "shut out the sunshine."[2]

The numbers that would come together to feed and roost exceeded what seemed possible. Puritan clergyman Cotton Mather writes of a pigeon roost established in December, an unusual time of year for the Salem, Massachusetts, area. Their presence was explained by a large crop of acorns left uncovered by snow. "At their lighting on a place of thick Woods, the Front wheel'd about, the Flanks wheel'd inward, and Rear came up and pitch'd as near to the Center, as they could find any Limb, or Twig, or Bush to seize upon. Yea, they satt upon one another like Bees, till a Limb of a Tree would seem almost as big as an House."[3]

Mather sent two passenger pigeon papers to the Royal Society of London, of which he was the first member from the United States. While he was the first writer to discuss pigeon milk, the substance fed to squabs that enables them to grow so quickly (information he received from local tribes), he also believed the birds migrated "to some undiscovered Satellite, accompanying the Earth at a near distance." The reason, he believed, that pigeon nests were so loosely constructed was to cool off the eggs, which, given the unusually high temperature that he attributed to them, would otherwise burn up. Perhaps because of such views as the last two, the Royal Society never published Mather's papers in full.[4]

John Josselyn, visiting New England in 1638 and 1663, provided his take: "I have seen a flight of Pidgeons in the spring and at Michaelmas when they return back to the Southward for four or five miles, that to my thinking had neither beginning nor ending, length, or breadth, and so thick I could see no sun." Around 1659, in September, the Reverend Andrew Bernaby was leaving Newport, Rhode Island, when the pigeons caught his eye: "I observed prodigious flights of wild pigeons: they directed their course southward, and the hemisphere was never entirely free from them. They are birds of passage, of beautiful plumage, and are excellent eating. The accounts given of their numbers are almost incredible."[5]

In fact, writing a hundred years later about Florida, William Stork found passenger pigeon numbers so incredible, he refrained from elaborating, apparently feeling no one would believe him anyway. Such restraint was perhaps justified, for the annals tell of a Captain Davy, who was in Philadelphia in the mid-1700s when a huge flight of pigeons took hours to cross over the city. At some point thereafter he went to Ireland and talked of what he had seen. His listeners were so incredulous they called him a "whopping liar" and referred to him ever after as "Captain Pigeon."[6]

What the Irish dismissed as malarkey, some of those North Americans who had witnessed pigeon flights with their own eyes went in the other direction: they attached to the unusual numbers meanings, and meanings that went well beyond the prosaic lessons of natural history. They sought signs that would predict the future, and generally with regards to pigeon flocks, the divinations spelled trouble. In 1675 Virginians saw "three prodigies in that country, which, from th' attending Disasters, were Look'd upon as ominous presages." The first was a comet that moved through the sky every night for a week; the second involved "Swarms of Flyes about an inch long . . . rising out of Spigot Holes in the Earth," which were possibly periodic cicadas. And then there was the third: "Flights of pigeons in breath nigh a Quarter of the Mid-Hemisphere, and of their Length was no visible Ends; Whose Weights brake down the Limbs of Large Trees whereon those rested at Nights; This Sight put the old planters under the more portentous Apprehensions, because the like was Seen . . . in the year [1644] When th' Indians Committed the last Massacre." To many, these omens seemed to be fulfilled a year later with the outbreak of Bacon's Rebellion, whereby a group of Virginians led by Nathaniel Bacon first defeated marauding Indians and then eventually became so furious with the sitting governor—he had called them traitors—they drove him from office and burned down the capital, Jamestown.[7]

Another series of large flights over Philadelphia in the hot summer of 1793 also evoked fear in many, as they saw it as a sign of bad air and forthcoming evil. Sure enough, beginning in August, the city was seized by a yellow fever outbreak, one of the worst epidemics in American history. Of a population of fifty-five thousand, between four and five thousand died, and many fled, leaving the city struggling to survive.[8] Henry Wadsworth Longfellow immortalized the events in his classic poem of lost love, *Evangeline*:

Then it came to pass that a pestilence fell on the city,
Presaged by wondrous signs, and mostly by flocks of wild pigeons,
Darkening the sun in their flight, with naught in their craws
but an acorn.

The most familiar accounts of passenger pigeons belong to the preeminent bird students of the early nineteenth century, Alexander Wilson, called the father of American ornithology, and John James Audubon, America's most famous ornithologist. The bulk of their observations take place in Kentucky, Wilson's in 1810 and Audubon's in 1813. If you have read anything about the species, you have probably read an excerpt of varying length from one or the other or both. William French in his *Passenger Pigeons in Pennsylvania* writes that these two ornithologists did such an admirable job describing nesting colonies that no one thought it worthwhile to try again. Fortunately, that is not true, but such is the stature of their work.

A poet in his native Scotland, Wilson's skills with language come through in his portrayal of the birds in flight:

> The appearance of large detached bodies of them in the air, and the various evolutions they display, are strikingly picturesque and interesting. In descending the Ohio by myself, in the month of February, I often rested on my oars to contemplate their aerial maneuvers. A column, eight or ten miles in length, would appear from Kentucky, high in air, steering across to Indiana. The leaders of this great body would sometimes gradually vary their course, until it formed a large bend, of more than a mile in diameter, those behind tracing the exact route of their predecessors. This would continue sometimes long after both extremities were beyond the reach of sight; so that the whole, with its glittery undulations, marked a space on the face of the heavens resembling the windings of a vast and majestic river.[9]

As difficult as it is for people of today to imagine the kinds of numbers evoked by these old accounts, it is even harder to conjure up the power represented by birds that travel in groups of hundreds of millions or more. Wilson was on the river on another occasion and paddled to shore to buy some milk from a farmer. As he stood inside the cabin chatting, an amazing thing happened: "I was suddenly struck with astonishment at a loud rushing roar, succeeded by instant darkness, which, on the first moment, I took for a tornado, about to overwhelm the house and everything around in destruction." But his companions remained cool and calmly replied, "It's only the Pigeons."[10]

Audubon was gifted not only in his ability to create memorable pictures in paint, but in prose as well. Here he describes how a flock prepares to alight:

> As soon as the Pigeons discover a sufficiency of food to entice them to alight, they fly around in circles, reviewing the country below. During their evolutions, on such occasions, the dense mass which they form exhibits a beautiful appearance, as it changes its direction, now displaying a glistening sheet of azure, when the backs of the birds come simultaneously into view, and anon, suddenly presenting a mass of rich deep purple. They then pass lower, over the woods, and for a moment are lost among the foliage, but again emerge, and are seen gliding aloft. They now alight, but the next moment, as if suddenly alarmed, they take to wing, producing by the flapping of their wings a noise like the roar of distant thunder, and sweep through the forests to see if danger is near.[11]

Many who witnessed the passenger pigeon hordes refer to the hours when the sun was blocked by the bodies of the birds. Audubon seems to provide the most detailed account of an instance when massive flocks created a dusk lasting for days. In 1813, Audubon resided in Henderson, Kentucky, a town rising from the banks of the Ohio River. On one particular fall day he embarked on a trip to Louisville, 122 miles away. Just on the other side of Hardinsburg, about halfway to his destination, the pigeons began flying "in greater numbers than I thought I had ever seen them before . . . I traveled on, and still met more the farther I proceeded. The air was literally filled with pigeons; the light of noonday was obscured as by an eclipse; the dung fell in spots, not unlike melting flakes of snow; and the continued buzz of wings had a tendency to lull my senses to repose." The volume of birds coursing southwest never abated over the hours it took him to reach Louisville by early evening. Nor did they for three days running.[12]

Audubon and Wilson had a feud dating back to March 19, 1810. Wilson was peddling subscriptions for his multivolume masterpiece, *American Ornithology*, when he visited Louisville with a "letter of recommendation" to Audubon. He later wrote about that visit: "[I] neither received one act of civility from those to whom I was recommended . . . Science or literature has not one friend in this place." Years after Wilson had died, Audubon provided his own account of the encounter: He generously shared his own paintings with Wilson and even invited the visitor on field outings and to dinners with friends. When Wilson left town, he didn't even say good-bye to Audubon or the others who had welcomed him.[13]

Wilson was the older man and had started his monumental project to illustrate all the birds of North America before Audubon did. Although Audubon was by far the more accomplished painter, Wilson was the better scientist. This conclusion is bolstered by what each man wrote in response to passenger pigeons. Wilson calculated the size of one huge flight he witnessed as containing 2,230,272,000 birds. Audubon's figure for a large flight that he saw some years later totaled 1,115,136,000. It is not credible that Audubon independently arrived at an amount that was exactly one half of Wilson's.[14]

On another point, Wilson wrote that the pigeons laid one egg, while Audubon placed it at two. Wilson was surely correct, although each was apparently relying on secondhand information. And even Audubon's gorgeous drawing of the species mistakenly depicts a female perched on a branch passing food to the male on a lower branch. In reality they would have been next to each other, and the female would have been the recipient of the food.[15]

John Audubon's connection to the pigeons may have extended even beyond his own death in 1851. A strong tempest hit the Hudson River valley one fall night in 1876. Struggling in its clutches, a flock of tousled passenger pigeons finally found respite in the trees overlooking Audubon's grave site in the Trinity Church Cemetery in upper Manhattan, between Broadway and Riverside Drive. They stayed the night and most of the next day, foraging on the broad lawn as they sought to recover from their rough flight. Gardeners attending to the grounds in the morning swept clear the leaves, pigeons, and other debris deposited by the winds. Finally, come evening, most of the pigeons lifted off and headed out across the river in an elongated string. Those few that stayed soon became fodder for wandering cats. In concluding the story, John French asks if "in the eternal verity of things . . . some spiritual compass drew these storm-tossed and much persecuted birds toward the then unmarked resting place of their friend . . . and there found surcease for their sorrows"? Well, the answer is "of course not," but the question is a pretty one, and if the account is true (no source is given), it provides a touching postscript to the relationship between America's most celebrated ornithologist and its most remarkable bird.[16]

Taking the mantle from Audubon and Wilson as the species' premier

Simon Pokagon. Wikimedia Commons

observer, Simon Pokagon studied the bird in various parts of the Midwest during the critical period from 1840 to 1880. Producing language every bit as vivid and important as his predecessors, Pokagon was one of the most extraordinary commentators to have contributed to the passenger pigeon literature. He was the last chief of the Pokagon Band of Pottawatomie, a group who once held dominion over much of northern Illinois, southern Wisconsin, northwest Indiana, and southwest Michigan. In 1833, his father sold the band's holdings to the federal government for three cents an acre. Despite having consummated the transaction on terms so favorable to itself, the federal government withheld full payment, and it took Pokagon decades to recover the money that was owed. Pokagaon kept a foot in both the white and Indian cultures and was an impassioned advocate for the rights of his people. His effectiveness was enhanced by his superb abilities as an orator and writer.

Some of the loveliest and most instructive words devoted to the passenger pigeon came from his hand:

> It was proverbial with our father that if the Great Spirit in His wisdom could have created a more elegant bird in plumage, form, and movement, He never did. When a young man, I have stood for hours admiring the movements of these birds. I have seen them fly in unbroken lines from the horizon, one line succeeding another from morning until night, moving their unbroken columns like an army of trained soldiers pushing to the front, while detached bodies of these birds appeared in different parts of the

> heavens, pressing forward in haste like raw recruits preparing for battle. At other times I have seen them move in one unbroken column for hours across the sky, like some great river, ever varying in hue; and as the mighty stream, sweeping on at sixty miles an hour, reached some deep valley, it would pour its living mass headlong down hundreds of feet, sounding as though a whirlwind was abroad in the land. I have stood by the grandest waterfall of America and regarded the descending torrents in wonder and astonishment, yet never have my astonishment, wonder, and admiration been so stirred as when I have witnessed these birds drop from their course like meteors from heaven.[17]

CASTING DEEP SHADOWS

> Every afternoon [the pigeons] came sweeping across the lawn, positively in clouds, and with a swiftness and softness of winged motion, more beautiful than anything of the kind I ever knew. Had I been a musician, such as Mendelssohn, I felt that I could have improvised a music quite peculiar, from the sound they made, which should have indicated all the beauty over which their wings bore them.
>
> —MARGARET FULLER, ON THE ROCK RIVER
> NEAR OREGON, ILLINOIS, 1843

Unlike such natural spectacles as the geysers of Yellowstone or the herds of bison grazing across the rolling grasslands of the Great Plains, one did not have to travel to remote districts to see passenger pigeons. In their movements across the eastern half of the continent, these birds cast deep shadows over Boston, New York, Philadelphia, Montreal, Toronto, Cleveland, Columbus, Detroit, St. Louis, Minneapolis, Chicago, Louisville, and other cities. The pigeons did not appear everywhere every year, and their abundance ebbed over the decades from east to west as the forests upon which they depended gave way to agriculture and economic development. But as long as there were large flocks, most anyone had the chance to experience the intense emotions evoked by this bird. For most of the Midwestern cities, this chance lasted through the late 1860s and early to mid 1870s.

Philip Hone, mayor of New York City from 1826 to 1827, assiduously maintained a diary for thirty-one years, leaving to future generations a detailed glimpse of that time and place. Through reading and listening to friends, he had long been familiar with the aerial splendor of passenger pigeons on the move, but he did not get to see the sight himself until November 4, 1835, while in Mattawan, New York: "They came from the west, and crossing the valley where I was, passed the top of the mountains and went to the south and east. The air was filled with them; their undulation was like the long waves of the ocean in a calm, and the fluttering of their wings made a noise like the crackling of a fire among dry leaves or thorns. Sometimes the mighty army was scarcely visible in the bright blue sky, and in an instant a descent of astonishing rapidity brought them so low that if we had been provided with guns, it would have been literally 'every shot a pigeon.' " He was pleased to have finally observed the spectacle, in part because he would thereafter be able "to talk 'pigeon' with Audubon, in his own language."[18]

The residents of Columbus, Ohio, reacted to a large flight of passenger pigeons not with wonder but fear. One warm spring morning in 1855 the people of that city were going about their usual routines when they first noticed "a low-pitched hum" that slowly engulfed them. It grew louder, as horses and dogs began fidgeting. Then just within the limits of vision, wispy clouds appeared on the southern horizon: "As the watchers stared, the hum increased to a mighty throbbing. Now everyone was out of the houses and stores, looking apprehensively at the growing cloud, which was blotting out the rays of the sun. Children screamed and ran for home. Women gathered their long skirts and hurried for the shelter of stores. Horses bolted. A few people mumbled frightened words about the approach of the millennium, and several dropped on their knees and prayed . . . Suddenly a great cry arose from the south end of High Street. 'It's the passenger pigeons! It's the pigeons!' . . . And then the dark cloud was over the city . . . Day was turned to dusk. The thunder of wings made shouting necessary for human communication." When the flock had finally passed almost two hours later, the town looked ghostly in the now-bright sunlight that illuminated a world plated with pigeon ejecta.[19]

The many pigeons that flew over Cleveland in early March of 1860 provided an opportunity to perform a unique and cruel experiment. The owner of one of the local fireworks houses decided to see how the birds would react to hissing skyrockets in their midst. He launched several heavy missiles into a large group of birds, causing the flock to divide and scatter in various directions. Some landed and wandered about in seeming panic. The explosion of one projectile just below a flock caused the birds to rise until they were no longer visible. Many of the birds that opted for the ground were caught by boys. No mention is made that the rockets caused any direct avian fatalities, but all those present "enjoyed the sport as peculiarly original and well worthy the Spirit of the Times."[20]

Chicago is a good place to look at the pigeon flights over time. Being on the southern end of the only Great Lake on a north/south axis made it a particularly advantageous location to observe bird migration. Over the nineteenth century, the city grew at a rate matched by few if any others. From 1840 to 1870, the population increased from 4,470 residents to 298,977. By 1890, it would exceed a million.

A newspaper story from September 17, 1836, reported that within the past several days "our town was swarming with pigeons, the horizon in almost every direction was black with them." Nineteen years after that, yet another article claimed "a flock of pigeons, over six miles in length," crossed the city's skies. The species was still considered a "very abundant" migrant and nester in small numbers up to 1876. Another observer gives May and June of 1881 as the last time "they were at all abundant in Cook County," where the city is located.[21]

The last big pigeon flight I know of in the region appeared in the spring of 1871 over the South Shore Country Club, a marshy area that back then was just southeast of Chicago, as it had not yet been annexed by the city. A hunting party arriving there that spring learned that ducks were largely absent but the jacksnipe were plentiful. In less than an hour they had bagged "as many birds as the right kind hunters care to kill." After a leisurely lunch of roast snipe and ample libations, the men headed back. The driver suddenly pulled to a stop and pointed to a dark cloud heading quickly toward them. One of his passengers readily identified the cloud as wild pigeons and exhorted the driver to accelerate so they could be close enough to do some unexpected shooting. But it was not to be, as when the flock spied Lake Michigan, it headed east and away from the hunters.[22]

According to Henry Eenigenburg, who lived next to the Calumet marshes on Chicago's southeast side, the fall of 1871 marked the end of the passenger pigeons as a common nesting species at the south end of Lake Michigan. The birds, he said, used to nest in the white pines that were still common in the Indiana Dunes. (These were clearly not in the huge nestings that occurred elsewhere but in the smaller configurations that few observers seem to have described in much detail.) But virtually no rain fell that summer, so that by September the entire region flanking Lake Michigan was a tinderbox. On October 8 the flames erupted and burned for several days. Peshtigo, Wisconsin, lost 1,152 people, and four square miles of Chicago became ash and rubble. Eenigenburg claims millions of pigeons also perished, which is doubtful, but the habitat that attracted them did suffer. Perhaps worse than the impacts of the fire itself was that little standing timber, particularly the highly coveted white pine, would survive the rebuilding of Chicago.[23]

RARE BIRDS: THOSE WHO PROTECTED AND APPRECIATED PASSENGER PIGEONS

> If the laudable quest for survivors of the species proves not forlorn, we trust our boasted humanity will hold the protection of this beautiful bird to be a most sacred trust—an attitude rarely taken in the day of its abundance.
>
> —ALBERT HAZEN WRIGHT, 1911

In the extensive passenger pigeon literature prior to the 1870s, many marveled at their numbers and movements and admired their beauty, although these sentiments were often uttered as the author participated in helping bring about the bird's extinction. Most statements of affection or appreciation were published only after the pigeons had disappeared. Although people kept live passenger pigeons as food, targets, and flapping decoys, for a bird so abundant, hearty, and innocuous (at least as individuals), there are surprisingly few mentions of them as pets. But the exceptions do exist, and they stand out as tiny islands in the sea of carnage.

Thomas McKenney was a Quaker, a pious man whose career became entwined with the treatment of Native people. Although he lamented that they were not being treated with the humaneness and justice they deserved, he also supported President Jackson's brutal Indian-removal policy, which with respect to the Cherokee even countermanded a Supreme Court decision. Despite McKenney's inconsistencies, his impulse toward the humane seems to have been stronger than that of most and stoked by his faith.

During the summer of 1826, McKenney left Washington, D.C., to become part of a delegation that negotiated the Treaty of Fond du Lac, then a post of the American Fur Company that would later become Duluth, Minnesota. He and his party of forty-three met with the Chippewa in a successful effort to get them to open some of their land to mining. As the armada of voyageur canoes headed back from Fond du Lac, they broke into small groups, traveling at different speeds and taking slightly different routes. At six in the morning of August 14, McKenney's boat and its companion were within sight of Keweenaw Point. The wind strengthened, Lake Superior foamed, and the crews struggled to keep their vessels on course. At that moment, a solitary bird, laboring to stay above the chop, was seen first heading toward the other canoe and then began following McKenney's. With a final burst of energy, it reached the upper yard, where it landed. As one of the paddlers raised his oar to strike the waif, McKenney grabbed his arm to stop the blow.[24]

The exhausted bird was handed to McKenney: "It was too feeble to fly. Its heart beat as if it would break. I took some water from the lake with my hand, into my mouth, put the bill of the little wanderer there, and it drank as much as would have filled a table spoon . . . It seemed to have sought my protection, and nothing shall cause me to abandon it." He looked about for a suitable container for the bird and placed it in a mocock (a kind of box) that had been given to him as a gift. He went on to speculate: "This is a member of the dove family, and the 'travelled dove' of the voyage. Is it a messenger of peace?—Why did it pass one canoe, and turn and follow another? Why come to me? None of these questions can be answered. But of one thing this poor pigeon is sure—and that is, of my *protection*; and though only a pigeon, it came to me in distress, and if it be its pleasure, we will never part." A few hours later, the party stopped to rest. McKenney noticed three Indians pounding corn between two rocks. They accepted his offer of tobacco for some of the corn, which he then fed to the bird, which gulped it eagerly.

The published account of this trip appeared a year later and included this touching footnote: "The pigeon called by the Chippeways *Me-me*, and by which name, it is called, is yet with its preserver—tame, and in all respects domesticated. It knows its name, and will come when called."

Another writer, identified only as F., tells of his journey through the Great Lakes in July of 1847 on the vessel *St. Louis.* The passengers comprised a distinguished group that included writers, editors, politicians, and clergy. They departed Chicago on July 7 bound for Green Bay, as they lazily made their way toward Buffalo. After stops in Milwaukee and Sheboygan, the boat continued northward until they reached the first opening to Green Bay. Dawn was breaking, the sun tracing wispy clouds in polished gold: "Everything around us was so calm, so bright so peaceful." The aura of tranquillity was suddenly enhanced by the appearance of a bird that symbolized peace like no other: "Peace's chosen emblem, with an arrow's speed, flew over us, and alighted not on a lovely lady's bosom, but on one of the iron rods extended between the smoke-pipes."[25]

F. waxed lovingly on this dove, not a "lumpish ungraceful" domestic rock pigeon, but "something far prettier: —a blue, free, fleet wild pigeon—a thing like Cora, untameable, and given to wild flights, but of a truly gentle disposition." Everyone on board took pleasure in the presence of the tired bird, hailing its appearance as a happy omen. John Smith, a fellow voyager familiar with the Great Lakes, explained that birds of all kinds were often found floating on the deep waters of Lake Michigan, and on occasion gales and deep fog claimed even passenger pigeons, otherwise noted for their speed and endurance.

As the passenger pigeon maintained its post, concern mounted among the observers that the heat of the pipes would prove untenable for the bird. Its drooping wings and an open bill eventually drove a sympathetic editor to grab a fishing pole; he wished to save the visitor by dislodging it from its torrid perch. Smith reached out and stopped him, however, pointing out that fatigue posed a greater threat to the bird than the heat. If it was otherwise, the pigeon would surely have left on its own. Sure enough, with the elapsing of half an hour and the increasing proximity to shore, the bird "launched into the air and sought the pleasant green-wood shade."

From Pennsylvania come tales of property owners objecting to the felling of their timber to get at passenger pigeons. They had no interest in the birds, however, but were merely attempting to halt the wanton destruction of their valuable property. Unique in the passenger pigeon annals is the short memoir written by Richard W. Wharton of Joaquin, Texas. His maternal great-grandfather John Clinton Payne immigrated to the United States from England in 1841. He wanted to see the frontier so he moved west from Virginia and settled down near Shelbyville, Texas, three years later. Through cabinetmaking and farming, he eventually amassed holdings in excess of seven hundred acres, some of which was old-growth forest and some an open area called the Old Prairie.[26]

While growing up in England, Payne developed a lifelong affection for birds. Fortunately, Payne's property held extensive tracts of maple and water oak (probably *Quercus nigra*), which drew large flocks of passenger pigeons in the fall. Unlike most landowners, he aggressively protected the pigeons on his land: "On one occasion, he caught two poachers with a sack of pigeons. With the three hired hands with him, they surrounded the poachers. My great-grandfather confiscated the birds, gave the two poachers a brief, intense sermon on the evils of poaching, issued them a diploma along with a few bruises."

The experiences of Joseph Dodson of Kankakee, Illinois, also stand alone in the relations of passenger pigeons and humans. Dodson became known for the birdhouses he built and sold, including his ninety-room, 490-pound purple-martin mansion. In his eighties, he wrote his recollections of growing up in Alton, Illinois, where the Illinois River joins the Mississippi. Pigeons by the millions streamed over their house, and hunters killed them in tremendous quantities. The Dodson family, repelled by the slaughter, sent young Joseph out with a small basket to collect injured birds. He searched the thick grass and shrubs where crippled pigeons had eluded the killers. Joseph returned home with as many as he could carry. His parents had built a wire coop to hold the birds while they mended. Over time, the Dodsons became pioneering wildlife rehabbers, gaining proficiency in repairing wings and legs. If a bird lacked both eyes or had both wings or both legs broken, they would have no choice but to euthanize it. But often injured birds would recover and be returned to the sky over which they were masters.[27]

A far more famous writer also had parents who would never kill passenger pigeons. Gene Stratton-Porter was one of the most popular novelists during the early 1900s. Her bread and butter were maudlin novels such as *Freckles* and *A Girl of the Limberlost.* But she also very much wanted to write natural history, which her publishers permitted from time to time to keep her happy. Some of these books include *Moths of the Limberlost* and *What I Have Done with Birds.* Much of her work was set in the Limberlost, a thirteen-thousand-acre wetland in northeastern Indiana where she believed she saw a passenger pigeon in 1912. But of greatest concern here is what she wrote of her parents. Her father was a farmer and a Methodist minister who hunted quail but felt more solicitous toward the pigeons for religious reasons, probably not unlike Thomas McKenney. The Strattons had twelve children, whom they sternly admonished to never shoot at either of the two native doves: "He used to tell me that they were among the very oldest birds in the history of the world . . . and he explained how the doves and the wild pigeons were used as a sacrifice to the Almighty, while every line of the Bible concerning these birds, many of them exquisitely poetical, was on his tongue's tip."[28]

One time Stratton-Porter visited some neighbors who were in the midst of dressing freshly killed pigeons: "I was shocked and horrified to see dozens of these beautiful birds, perhaps half of them still alive, struggling about with broken wings, backs, and legs, waiting to be skinned, split down the back, and dropped into the pot-pie kettle. I went home with a story that sickened me." Her father once again renewed his prohibition against any family member's shooting any dove. To his theological concerns, he added

Gene Stratton-Porter. Courtesy Indiana State Museum

the very material warning that if there was no cessation in the killing, the birds might disappear. Stratton-Porter acknowledged this was merely a precaution, for "that such a thing could happen in our own day as that the last of these beautiful birds might be exterminated, no one seriously dreamed."

James Fenimore Cooper, often considered America's first significant novelist, was the also the earliest writer to articulate "an American environmental conscience." Biographer Wayne Franklin calls Cooper's account of the passenger pigeon slaughter a defining moment, his initial call "to his fellow citizens—and the world—to imagine a better way of being on the earth." His daughter Susan Fenimore Cooper was one who did heed the call, as she would distinguish herself by becoming the first American woman to publish a book on nature when her *Rural Hours* appeared in 1850. (She, too, wrote a bit on passenger pigeons.) Another was Henry David Thoreau, whose work was in large measure inspired by the writings of Cooper.[29]

Cooper's most famous novels are the Leatherstocking Tales, featuring the adventures of the frontiersman Natty Bumppo, aka Leatherstocking. *The*

Pioneers came out in 1823 and was the inaugural book in the series. It also addresses issues that underlie crucial parts of the passenger pigeon story. In the words of one scholar, "much that happens in [the novel] is related to conservation in the broad sense of man's wisdom or lack of it in his manipulation of nature."[30]

In the novel, early on a late-April morning in upstate New York, Elizabeth Temple, daughter of leading citizen Judge Temple, awakes to the chattering of purple martins as they fly about the small houses crafted for their use. As she listens, she hears the louder cries of Sheriff Richard Jones urging her to arise: "Awake! Awake! My fair lady . . . The heavens are alive with wild pigeons. You may look an hour before you can find a hole through which to get a peep at the sun."[31]

As the morning unfolds, townspeople scurry about procuring whatever weaponry they can find, "from the French ducking gun, with a barrel near six feet in length, to the common horseman's pistol." For those without firearms, there are arrows propelled by both longbows and short bows. Then a horse-drawn cannon from a war fought long before is brought to bear on the massing pigeons. It is loaded with duck shot and discharged into the clouds of pigeons: "So prodigious was the number of birds that the scattering fire of the guns . . . had no other effect than to break off small flocks from the immense masses that continued to dart along the valley. None pretended to collect the game, which lay scattered over the fields in such profusion as to cover the very ground with fluttering victims."

Various characters then discuss what they have just witnessed, and three different viewpoints emerge. One group supports the dual notion that nature's richness is here solely for humans to exploit in any way and to whatever extent they see fit, and that the richness is inexhaustible. Supporting this position is Jones and the woodcutter Billy Kirby. Judge Temple represents the second view, that natural resources should be conserved for the future. (Commonly held today, this perspective was rare in Cooper's time.) Natty Bumppo holds a third position. It is a combination of a hankering for the old days, before there were farms and settlements and lots of people who kill beyond their needs, with a belief that plants and animals were put here for human use but not gross waste. There is also a dollop of sentiment for the victims of wanton human depredation: "I wouldn't touch one of the harmless things that cover the ground here, looking up with their eyes on me, as if they only wanted tongues to say their thoughts." One hundred seventy years later, the debate between the first two perspectives rages still, and how it is resolved may well determine the future of life on this planet.

Junius Brutus Booth

By far the strangest example of sympathy for the bird was expressed by Junius Brutus Booth, one of the leading actors of the nineteenth century and even more famously the father of John Wilkes Booth. Junius Booth suffered from alcoholism and bouts of depression, so his homage to the passenger pigeon might represent more a manifestation of pathology than true sympathy. But it is also a fact beyond dispute that he possessed a remarkably deep affection for animals and nature. Booth, a vegetarian, maintained his remote Maryland farm as a refuge where all hunting was forbidden; not even reptiles could be harmed. He rambled for long periods through the forests of his estate, escaping the treacherous and discordant world of humanity, while gaining succor from the vitality of the land.[32] This connection helped sustain him in his perpetual battle with the demons that lurked deep in his psyche. So in the end it is difficult to say exactly what motivated him to write his letter in January 1834.

Booth was in Louisville for an acting engagement when he wrote a local Unitarian clergyman, James Freeman Clarke, to secure a grave site for a recently departed friend. Clarke visited Booth's hotel to discuss the request further and to provide any consoling that might be needed. When Clarke arrived, he found Booth reading to another man, but that third person remained mute throughout Clarke's stay. Clarke relates what happened: "I asked him if the death of his friend was sudden. 'Very,' he replied. 'Was he a relative?' 'Distant,' said he."[33]

Booth then changed the topic, suggesting that he entertain his guests by reading Coleridge's *Rime of the Ancient Mariner* and "Shelley's argument against the use of animal food." He expounded at some length on his view that it was "wrong to take the life of an animal for pleasure," eventually offering Scripture in support when Clarke admitted that he found Shelley unconvincing on this issue. After more discussion, Booth finally offered to show Clarke the deceased. Upon entering an adjacent room, Clarke was shocked to see that the object of Booth's sorrow was a bushel of passenger pigeons! "Booth knelt down by the side of the birds, and with evidence of sincere affliction began to mourn over them. He took them up in his hands tenderly, and pressed them to his heart. For a few moments he seemed to forget my presence. For this I was glad, for it gave me a little time to recover from my astonishment, and to consider rapidly what it might mean."

Clarke had no idea how to take this: Was it a hoax or a practical joke at his expense? He concluded that Booth deeply revered all life, a view he considered exaggerated but one worthy of respect, "as all sincere and religious convictions deserve to be treated." Earlier in Booth's stay, a large flight of pigeons had triggered the typical slaughter, and baskets filled with the birds occupied the stands of mongers throughout the city. Clarke quoted Booth: " 'You see,' said he, 'they're innocent victims of man's barbarity. I wish to testify, in some public way, against this wanton destruction of life. And I wish you to help me. Will you?' " Clarke declined, and when asked whether it was because he feared ridicule, he said that it was because he did not agree with Booth's views. Booth wound up doing something less public than what he had originally intended but striking nonetheless: he commissioned a coffin for the deceased pigeons and had it transported in a horse-drawn hearse to a cemetery a few miles outside the city. He paid respects daily to his feathered relations, mourning with heartfelt grief.

Although Clarke refused the assistance that Booth sought, he may have been as understanding toward Booth's wishes as anyone in Louisville: "I could not but feel a certain sympathy with his humanity. It was an error in a good direction. If an insanity, it was better than the cold, heartless sanity of most men." Indeed, it was that heartless sanity of most men that drove the pigeon from the ranks of the living.

Anthony Philip Heinrich, a composer and musician of such virtuosity that he was dubbed the "American Beethoven," was born in Bohemia. He was adopted by a wealthy merchant, who left him extensive property holdings and a thriving wholesale trading business. Heinrich threw himself into these various enterprises and traveled throughout Europe to advance their interests. His taste for travel prompted him to cross the Atlantic in 1805 to glimpse the United States. Although untrained, from an early age he also

Anthony Philip Heinrich. Wikimedia Commons

had an abiding passion for music, particularly the piano and the violin. His love for the latter became indelible when, on a trip to Malta, he purchased a Cremona violin, one of the most revered types of that instrument. He would never again be without it.[34]

Unfortunately, those early years of affluence soon withered into poverty: his business became tainted by the unscrupulousness of others, and, more important, the Napoleonic Wars sapped the economic vigor of Europe. Heinrich made one last-ditch effort to reverse his fortunes by expanding his trade to United States. It might have worked but for the continuing slide of Europe, which led to a devastating financial collapse in 1811. Thereafter, he would lead a life of economic struggle and devoted himself to his music. But he never lost his energy and optimism.

In 1816 he was in Philadelphia, then the nation's center for arts and sciences. He did not stay long as he was offered a paid position to direct the theater of Pittsburgh. The formal arts in America were just beginning, and the nascent efforts were not always impressive. One observer described Heinrich's new professional home this way: "Such a theater! It was the poorest apology for one I had then ever seen." After a brief tenure in Pittsburgh, Heinrich headed to Kentucky, which was then the cultural center of the west. Heinrich was to stay in the state for five years, performing, teaching, and composing—all to great acclaim. An important period in his life was the several months he lived alone in a small cabin near the Catholic village of Bardstown, south of Louisville. He communed with nature, played his violin, and began composing. He forged many lifelong friendships while in

Kentucky, including one with John Audubon, who lived in Henderson. In a dedication to two friends, he wrote, "These compositions . . . were drawn up in the wilds of America, where the minstrelsy of nature, the songsters of the air, next to other Virtuosos of the woods, have been my greatest inspirers of melody, harmony, and composition."[35]

Heinrich traveled back and forth to Europe, spending a lot of time in London. He was widely recognized for his undoubted talents, as, for example, he was the first American composer to be included in a European encyclopedia of musicians. But his works were undeniably unusual. Reviewers acknowledged Heinrich's genius, but as one British critic noted, the compositions "resemble nothing that was ever seen before, so unaccountably strange and odd is their construction."[36]

On one occasion, Heinrich played a piece to President John Tyler and a few others in a White House parlor: "The composer labored hard to give full effect to his weird production; his bald pate bobbed from side to side, and shone like a bubble on the surface of a calm lake." After a bit, the audience began fidgeting and Tyler interrupted, asking Heinrich to play instead "a good old Virginia reel." Mortified, the composer rolled up his music and stormed out furious, cursing the president in German.[37]

Heinrich's "supreme triumph" was the concert devoted to his works in Prague in 1857, when he was seventy-six years old. Three compositions were performed, the final and most successful being the "symphonic poem" *The Columbiad; or, Migration of the American Wild Passenger Pigeons*. The program notes list its nine movements plus an introduction: "Introduction. A Mysterious Woodland Scene, the assembling of the wild passenger pigeons in the 'far west' for their grand flight or migration; I. The flitting of birds and thunder-like flappings of a passing phalanx of American wild pigeons; II. The aerial armies alight on the primeval forest trees, which bend and crash beneath their weight . . . ; VII. The alarm of hunters' rifles startles the multitude. The wounded and dying birds sink tumultuously earthward; VIII. In brooding agitation, the columbines continue their flight, darkening the welkin as they utter their aerial requiem, but passing onward, ever onward to the goal of their nomadic wandering, the green savannas of the New World."[38]

Heinrich's biographer William Treat Upton says, "We cannot read it through without feeling its romantic power. The situations are admirably chosen and tersely, yet poetically, expressed."[39] *The Columbiad* was the perfect finale for this momentous concert. It had long been Heinrich's dream to have his works performed in the capital of the country of his birth, and by musicians accomplished enough to master the complexity of his pieces, something not possible in the United States. Yet the grandest number of that memorable afternoon of music was devoted to that New World endemic, the passenger pigeon.

Suffering from ill health and impoverishment, Heinrich died in New York City four years later and was buried near John Audubon. Though never a popular favorite, Heinrich was an American original who found inspiration in the natural history of his adopted land. And like the pigeons he described in music, he is not nearly as well remembered as his life and work surely warrant.

Lewis Cross never forgot the flights of pigeons that coursed through his youth. Cross was one of four sons born to a pioneering couple who left New York to settle near Spring Lake, Michigan, in the late 1860s. The land proved to be unyielding so the family channeled its entrepreneurial efforts into producing butter-tub hoops out of the black ash that grew abundantly in the adjacent lowlands. With twelve hands devoted to the task and thirteen years of labor, they saved enough money to buy a new homestead at Deremo Bayou on the Grand River, several miles away. Here the longer growing season and more fertile ground enabled them to raise a variety of fruit, and their holdings grew to nearly a hundred acres. Not much is known about the elder Crosses, but it speaks well of them that each of their boys attended college, the state Normal School at Valparaiso, Indiana, now Valparaiso University.[40]

Cross began painting as a child and, after experimenting with different media, decided that he preferred oils and to a lesser extent crayon and pencil. He felt that watercolors required too much accuracy. The size of his works ranged from the small to canvases exceeding ten feet in width and length. When Cross talked about his art, he always emphasized that it was a hobby: "I paint because I like to."[41] He never aggressively marketed or exhibited his work, but he sold hundreds of paintings, most, presumably, to buyers in the area. One newspaper article notes that some of his work was shown by the local women's club, which suggests that few in the larger world saw it.

Cross drew what he knew, mostly scenes depicting local history and the wildlife that used to abound. The best example of that was the passenger pigeon, of which he produced a number of paintings, all modeled after the single bird he shot and mounted decades earlier. He felt a sense of obligation to preserve a record of what he was privileged to have witnessed but was now gone forever: "There are not many of us left who remember the pigeons as they were then. Maybe some of my work is not artistic but it is historical . . . I can remember back to the 70s when the sky would be so filled with them that the sun would be obscured for as long as an hour. At other times, when the sun was in the right position, a flock would appear as a perfect rainbow, caused by their iridescent coloring."[42] One of his most striking passenger pigeon paintings and the actual stuffed bird on which they were based are well displayed in the Lakeshore Museum Center in Muskegon.

Like most of his contemporaries, he rejected the notion that humans alone could have wiped the species out, despite the killing he himself witnessed. He was more of the view that the birds were killed in a hurricane, perhaps, as they were headed to South America. But he did acknowledge that none of the reports of pigeons in new locations ever proved to be accurate.[43]

From 1910 to 1914, Cross designed and built himself a two-story mansion out of concrete blocks. Called the Castle, it was situated on family property overlooking Deremo Bayou. The papers pointed out that he never married and lived alone in his house. Perhaps had he lived today, he would have shared the house with another. He was not a hermit, however, as he offered art classes on the upper floor, which held his studio and a small gallery. Schoolchildren and high school art students were among the many visitors he welcomed. Cross did not smoke or drink, but attributed his good health and long life to work. Even as an octogenarian he was capable of painting his detailed oils without the aid of glasses. But when at last, at the age of eighty-eight, he could no longer live independently or perform the activities that had sustained him for so long, he took his twelve-gauge shotgun and entered the lost world of his subjects. Boys calling at his home for apples found his body at the foot of the stairs.[44]

CHAPTER 4

Pigeons as Provisions to Pigeons as Products

PIGEONS AS PROVISIONS: FOOD, FEATHERS, AND MEDICINE

When I can shoot my rifle clear,
To pigeons in the skies,
I'll bid farewell to pork and beans,
And live on pigeon pies.

—DITTY COMMON IN 1850S

Although Jacques Cartier was the first European to see passenger pigeons, he was not the first one known to have killed any. (Perhaps he and his crew were sated by the casks of great auks that they had collected and salted earlier.) As far as I know, that honor falls to the settlers of the short lived Fort Caroline, established by the French near where Jacksonville, Florida now stands. Sometime between January 25, 1565 and May 1565, Rene Laudonniere, the fort commander, reported that "a great flock of doves came to us, unexpectedly and for a period of about seven weeks, so that every day we shot more than two hundred of them in the woods around our fort." Twenty years later, Thomas Heriot implied the taking of passenger pigeons when he wrote that the area around Roanoke Island (North Carolina) "provided Turkie cocks and Turkie hennes, Stockdoves." And in 1605 Samuel de Champlain and his crew became the second known Europeans to have specifically stated that they preyed on the species when he tarried off southern Maine and encountered "countless numbers of pigeons, whereof we take a goodly quantity."[1]

The generations who came later would take ever more goodly amounts of pigeons. As amazing as was the abundance of the pigeons, the litany of slaughter dominates the history of this species. People killed them in virtually every way imaginable and for many reasons. And at times, seemingly for no reason at all.

The newly arrived Europeans looked at the masses of pigeons both with wonder and hunger. The new continent possessed fecundity beyond what they had ever seen, and the pigeons manifested the pullulation of life to the ultimate degree. They were easy to catch and were seemingly inexhaustible, albeit unavailable at any given location for months or even years at a time. But when they were in the vicinity, their presence provided a reliable source of food, the absence of which would have made some pioneering efforts even more difficult, if not impossible.

For example, a great crawling pestilence befell much of New Hampshire in 1781. The unidentified larvae "destroyed the principal grains that year," eliminating both bread stuffs and silage for cattle and pigs. The lack of food became so severe the residents of several newer settlements considered pulling up stakes and leaving. But they were rescued by a bumper crop of pumpkins in Haverhill and Newbury, two of the older towns, and the arrival "of an immense number of passenger pigeons." It would be nice to say that the pigeons arrived in the nick of time and devoured the offending arthropods like the California gulls that feasted on the locusts that plagued the Mormons decades later. But, in fact, the birds showed up "immediately upon the disappearance" of the insects. Although probably somewhat disappointed by the timing, the residents gave full credit to Providence nonetheless for sending the legions of pigeons in quantities that could not be exceeded, "unless [by] the worms which preceded them." The Tyler family of Piermont took advantage of the new arrivals and in the next ten days caught over four hundred dozen. They invited their neighbors over for a picking bee. The helpers kept the meat of all the birds they stripped, and the Tylers acquired enough feathers to make four good beds. Pigeon meat preserved for the winter proved to be a critically important replacement for the lost cereal.[2]

Pigeons were important to many early settlers. An official report on the status of the Niagara Colony in 1785 credited wild pigeons and fish as being the principal food that sustained the residents through the summer until crops could be harvested. All the Crown needed to supply was a small amount of flour to each person. At Galt, Ontario, in 1832, when the combination of a summer-long drought and an August frost wiped out most of the crop, the locals were saved from starvation by netting or shooting hundreds of pigeons.[3]

The residents of the young Plymouth Colony came to see the pigeons as a double-edged sword, or, in the words of John Winthrop, "the Lord showed us, that He could make the same creature, which formerly had been a great chastisement, now to become a great blessing." In 1643, unusually chilly and wet weather reduced the corn crop alarmingly. And to worsen the situation, just before harvest time, the pigeons arrived ("above 10,000 in one flock") and consumed most of what there was. Five years later, the pigeons made another appearance, but this time after the harvest was completed. In this instance, Winthrop proclaimed the birds "a great blessing, it being incredible what multitudes of them were killed daily."[4]

As the white presence in North America increased and no longer needed rescuing by hordes of passenger pigeons, the birds became less out-of-the-blue lifesavers and more a dietary mainstay. A traveler riding between Newport and Boston in September 1759 found that little other food was available and locals subsisted almost exclusively on pigeons. Masters fed servants pigeons with such frequency in places that the employees demanded clauses in their contracts restricting how often this sort of poultry was served. Alexander Wilson told of the times when passenger pigeons were not eaten merely once a day for days on end but were the starring ingredient for breakfast, lunch, and dinner over many days. At that point, he said, "the very name becomes sickening."[5]

The territory around Wells and Kennebunk, Maine, in the mid-1700s was rich in whortleberries, which drew the pigeons in "innumerable numbers." Gunning them down was an exercise in democracy, for the pastime attracted everyone, including wealthy men in their prime, the elderly, children, women, and slaves. An accomplished shooter could bring in three hundred in a day. But the only way one could give away the surplus was by offering the birds fully dressed.[6]

From the ruthless standpoint of the market, the ultimate manifestation of profligate slaughter is when the prey is killed in such quantities that the accumulated corpses are valueless. Two pence a dozen was the going price in Boston in August 1736, and many could not even be sold at that amount. About the same time, in Granby, Massachusetts, surplus birds were fed to pigs, a practice that was not uncommon at nestings and roosts. Of all the awful things that were done to passenger pigeons, the great passenger pigeon historian A. W. Schorger devoted a single sentence to this use of the bird: "The extensive feeding of pigeons to hogs is unworthy of comment." At least those who did this relieved the hunger of their livestock. Not even that much could be said for what J. Benwell observed on his trip from Cleveland to the Ohio River in what was probably the 1840s: "We saw many carcasses of these birds outside the villages, such numbers having been destroyed, that the inhabitants could not consume them, and they were accordingly thrown out as refuse."[7]

The problem of too many dead pigeons was solved slightly differently at a nesting in Pennsylvania. There the surplus was plowed into fields as fertilizer. In this area, a widely held belief was that burying dead passenger pigeons in the garden resulted in more colorful flowers.[8]

For those who could afford to be choosy about what they ate, questions arose as to the palatability of passenger pigeon flesh and how best to prepare it. G. W. Cunningham, writing in 1899, saw little culinary merit in the bird. He thought it was hardly better than a yellowhammer! Yellowhammer is an antiquated name for the northern flicker, a species of woodpecker that has a great fondness for ants. But the flicker did have its supporters, too, as expressed in that strange book of 1853 *The Market Assistant*, which is sort of a gustatory field guide to the birds. All the species that were stocked in the New York markets are discussed as to their suitability for the kitchen. Thomas DeVoe, the author, said that flickers were a seasonal specialty most often appearing in the fall, "when it is fat, and its flesh quite savory, but not so tender as the robin."[9]

As for passenger pigeons, DeVoe had decided ambivalence. Adult birds shot on the wing were "very indifferent eating, even if well and properly cooked." Much better were the adults caught live and fed grain in coops. But best of all were the squabs, most delectable "when fat and fresh." William Byrd also was not impressed with birds killed during migration, "though good enough upon the march, when hunger is the sauce, and makes it go down better than truffles and morels would do."[10]

A Kentucky author penned a memoir on passenger pigeons that eventually appeared in the *Indianapolis Star* newspaper. After two trials, he found the flesh "as tough as whit leather, about as juicy as a pith of a dried corn stalk, as digestible as rawhide and almost as hard to masticate as rubber." This prompted a stout defense of the bird by C. G. M'Neill, who pointed out that one could say the same for beef or chicken if all you had tasted was an old bull and a four-year-old rooster. Another big fan of the bird as food was Etta Wilson, who averred that she "never ate a pigeon of any age that was not delicate and delicious."[11]

Numerous dishes featured the birds. Pigeon pie was one favorite, and stewed pigeon another. An 1857 guide to help Canadian settlers offers this recipe: "To make a pot pie of them, line the bake-kettle with a good pie crust; lay out your birds, with a little butter on the breast of each, and a little pepper shaken over them, and pour in a tea cupful of water—do not fill your pan too full; lay in the crust, about half an inch thick, cover your lid with hot embers and put a few below. Keep your bake-kettle turned carefully, adding more hot coals to the top, till the crust is cooked."[12]

Two recipes for stewed pigeon, one from Madison and the other from Chicago, are similar. In one the birds are stuffed with finely chopped bread and pork, while the other adds hard-boiled eggs to the mixture. Bard the birds with pork, place them in a tightly covered kettle with enough water to cover them, and put them into the oven until done. The real difference is in the sauces that finish the dishes. The Chicago ladies preferred mixing the pigeon gravy with the juice of one lemon, a tablespoon of currant jelly, and enough flour to thicken the boiled sauce. Perhaps in homage to one of her state's principal industries, Mrs. Hobbins of Madison simply added butter and cream to her gravy.[13]

Although not especially fond of passenger pigeon, George Sears, writing as Messmuk, recognized that one might be out in the woods where there was little else to eat. To make the best of the situation he suggested boiling the birds until they were tender and then taking them from the pot. Remove the breast meat, dredge it in flour, and then panfry as you would squirrels. The giblets and rest of the carcass should be stewed for a later meal. A Canadian observer from the 1770s reported that the pigeons "furnish soups and fricassees, which are usually dressed with a cream sauce and small onions." One unusual preparation was contributed by M. W. Althouse of Toronto: "Hunters and maple sugar makers often cooked adult pigeons by roughly drawing and then enclosing the unplucked carcass in wet clay which was then covered with the hot embers and wood ashes. When cooked the meat was removed from the covering of baked clay which kept with it all the feathers and most of the skin of the bird."[14]

People used pigeon flesh in all kinds of ways. As food for long voyages, roasted pigeons were kept wholesome by cramming them into barrels where they were covered with melted lard and mutton fat that would congeal to form an airtight seal. Most surplus pigeon meat, though, was salted, smoked, or pickled. A fancy variation on the last technique was employed by the mother of Charles Belknap when they lived in Grand Rapids, Michigan. By her hand, numerous pigeons wound up in earthen jars, where they were preserved with spiced apple cider and served to special guests: "The minister never had to eat woodchuck in our house." A couple from Illinois preferred to age and cure their meat, as a guest discovered when he was led to their attic, where many hundreds of dried pigeon breasts hung from the rafters on hooks.[15]

From their pedestrian beginnings as food for fearless explorers and struggling pioneers, passenger pigeons followed their human predators up the social ladder to become important components of some of the fanciest meals served in nineteenth-century America. Charles Dickens had just turned thirty years old in February 1842, but his literary success, embodied by such novels as *Pickwick Papers*, *Oliver Twist*, and *Nicholas Nickleby*, earned him accolades everywhere he traveled on his tour of the United States. When New Yorkers had the chance to fete the celebrated author, they staged a banquet that was "the finest civic pride could produce." The meal was held in the venerated City Hotel, the grande dame of local hotels since it was built in 1793, and catered by one Mr. Gardener at the cost of $2,500. Sitting at the head chair was Washington Irving, just one of the many luminaries present. After working their way through the first course of soup and/or fish, and a second course of eighteen items (mostly meats labeled either cold, roasted, or boiled), they had the chance to explore nineteen more selections denoted as entrées. Sharing the spotlight with such dishes as Larded Sweet Bread with Sorrel and Stewed Terrapin were three pigeon creations: Stewed Pigeons with Peas, Stewed Pigeons with Mushrooms, and Pigeon Patties with Truffles.[16]

Still a part of the gastronomic life of New York's bourgeoisie, the pigeons soared to even loftier heights with the opening of what was arguably the country's first truly great restaurant, Delmonico's. When Rear Admiral Lessoffsky steered his fleet into New York Harbor in November of 1862, Secretary of State Seward wanted to show the administration's heartfelt appreciation, for Russia was then the only major European nation publicly supporting the Union cause. It was also meant to impress Britain and France, who had not yet chosen a side in the war. While the venue was the Academy of Music, the food was Delmonico's. As one of the many entrées, the Russian officers and their hosts could have enjoyed *côtelette de pigeons à la macédoine*, pigeon cutlets served with a medley of diced vegetables laced with butter. Lately Thomas comments that this meal demonstrates how "gastronomy can be made to serve two purposes simultaneously: in this case, to give delight to friends and to give potential enemies indigestion." Four years later, President Andrew Johnson came to visit. From City Hall, where he was hailed by the mayor, Johnson and his party (which included his successor, General Grant) formed a parade that moved on to Delmonico's through throngs of cheering New Yorkers. Each of the dinner courses was paired with a wine, and for the entrées the selected libation was Château Margaux '48, which presumably went well with each of the six offerings, including *ballotines de pigeons Lucullus*, an amazingly rich dish of boned pigeon stuffed with foie gras and truffles coated in aspic and garnished with cock's combs, cock's kidneys, and more truffles.[17]

Meat was not the only part of the bird that was valued. Fat and feathers were coveted as well. The fat was used as shortening and even in the making of soap. Two visitors to a large nesting on the Susquehanna River in northern Pennsylvania in 1810 said that millions of chubby squabs were reduced to oil, which was then packed in barrels and sent downstream on boats. Generally feathers were collected from birds killed for the meat, but sometimes the feathers were the primary product and the carcasses were discarded or fed to swine. Most of those who picked the feathers were women and children. In Coudersport, Pennsylvania, after one particularly large nesting, mother pluckers were hired at the rate of five cents for every dozen pigeons they processed. The greatest reward to the participants, though, was social, as it took on the air of a picnic where folks could gossip while earning a little money.[18]

Beds and pillows stuffed with pigeon feathers enjoyed broad popularity. In the early years of Saint-Jérôme, Quebec, an acceptable wedding dowry had to include a pigeon mattress and pillows. Many people in various parts of pigeon range believed that these sleep articles provided eternal life. One critically ill lady in Ontario was moved from her bed of pigeon feathers to one of more mundane stuffing to hasten her end. In 1936, Alvin McKnight of Augusta, Wisconsin, related how he and his wife continued to sleep on a bed that they received in 1877 filled with the feathers of 144 dozen passenger pigeons. They were both in excellent health despite his age of eighty-four and hers of seventy-seven. They were beginning to believe that such beds would repel death and were prepared to test the proposition. The pigeon feathers, however, were no more able to grant immortality to the McKnights than they were to the species itself.[19]

In the eyes of some, passenger pigeon parts held one more valuable property: medicinal. Dr. John Brickell, writing on the natural history of North Carolina in 1737, stated that the blood was effective in the treatment of the eyes and, when swallowed, "cures bloody fluxes." He also had a good word for the dung, saying it could relieve most anything that ails, including headaches, pleurisy, apoplexy, and lethargy. How the physician administered the dung is left obscure. A Native healer from Quebec, on the other hand, treasured the gizzards, stringing them up to dry so she could use them to treat gallstones. The logic here was that the pigeons would at times ingest small stones, yet suffered no ill effect because the gizzard had the power to dissolve them. Therefore, if ingested, the gizzard would come in contact with the patients' stones and make them likewise disappear.[20]

THEY MADE GREAT HAVOC: ENEMIES OF AGRICULTURE

Even in areas not yet converted to crops, some landowners feared that the pigeons would jeopardize their timber holdings and took measures to keep the birds away. The *Niles (MI) Republican* of April 25, 1850, published this singular report by a local settler: "I am completely warn down. The pigeons are roosting throughout our woods and the roost extends for miles. Our neighbors and ourselves have for several days had to build large fires and keep up reports of fire arms to scare them off. While I write, within a quarter of a mile, there are thirty guns firing. The pigeons come in such large quantities to destroy a great deal of timber, break limbs off of large trees, and even tear up some by the roots."

Alexander Wilson calculated that two billion passenger pigeons would consume almost 17.5 million bushels of mast a day. The vast amount of food that the species would need to sustain itself prompted one British ornithologist to wonder why "any farmer should ever dare to migrate to America." The literature is mixed on whether the pigeons were a serious threat to agriculture. Farms in proximity to natural forage often escaped pigeon depredations. This would seem to be the case, for example, in the Wilderness District of Nicholas County, West Virginia, where "a vast multitude" of birds roosted in the fall of 1876. It has been suggested that heavy baiting by hunters at a nesting in New York might also have relieved the pressure on farmers. (Given the numbers of birds and how much they ate, this scenario becomes plausible only when accompanied by pigeon killing on a fantastic scale.) But things were apt to be different where hungry birds found themselves remote from native food sources. Then it was open warfare.[21]

A Jesuit traveling up the St. Lawrence River to Montreal during 1662–63 commented on the huge numbers of pigeons: "This season they attacked the grain fields where they made great havoc, after stripping the woods and fields of strawberries and raspberries which grow here everywhere under foot." But it was at a great cost to the pigeons, which were killed in such abundance that many corpses remained even after home consumption and the provision of servants. This surplus was either salted in barrels for winter use or given to dogs and pigs. Not long after, Baron de La Hontan reported, probably facetiously, that the pigeons were so loathsome in Canada, the bishop "has been forced to excommunicate 'em oftener than once, upon the account of the Damage they do to the Product of the Earth."[22]

Du Page County, Illinois, is just west of Chicago. It is today a place of suburban sprawl (which would have to include the older subdivision where I live), broken up by twenty-five thousand acres of forest-preserve land, plus assorted other tracts of open space. In February of 1852, the pigeons, called "pernicious varmints," arrived in Du Page and harvested every seed many unfortunate farmers planted. An article describing the events compares blackbirds with passenger pigeons. They conclude that blackbirds, "as evil and dark-hearted as they are," are still preferable to the pigeons.[23]

Pigeons in southwestern Michigan would descend on the corn and wheat fields to consume acres of produce. In Pennsylvania the birds were particularly partial to buckwheat. So many pigeons foraged on grain in Eden, Wisconsin, in 1869 they forced the farmers to abandon their fields to seek shelter in their homes. The farmers of western Iowa had no better success in their struggles against the pigeons in the mid-1860s. A detailed account of their woes refers to the pigeons as "a perfect scourge [that] lit upon the fields of new-sown grain, and rolling over and over like the waves of the sea, picked up every kernel of grain in sight." So the farmers tried again, but before they could harrow the freshly seeded soil, the pigeons would already have worked the ground. Some farmers were forced to try yet a third time. Nothing they did could keep the birds away, be it the shooting of guns, throwing of rocks, shouting, running to drive them off, barking dogs, or even killing vast numbers with poles.[24]

Children often had the job of keeping the pigeons at bay. While dad sowed, the kids would be dispatched to watch for pigeon flocks. Armed with such implements of noise as cowbells and metal pans, they would chase the birds until the seeds could be covered. But they had to be quick about it, as R. D. Goss discovered. He grew up in Wabasha County, Minnesota, during the 1860s. Back then wheat was sown by hand, so his father had taken advantage of a calm evening and morning to seed five acres. Goss had finished his chores and was headed to the barn to tend the horses when he spotted the pigeons. A cloud of them landed on the newly seeded field, and in that familiar wavelike action they began to feed. Goss ran toward the birds as fast as he could and managed to flush them. But to no avail: despite the short duration of their visit, the flock had cleaned three acres of the four and half bushels of seed that Goss's father had just planted. The pair could not find a single kernel of wheat.[25]

When Per Kalm made his trip through the eastern United States and Canada (1748–51), he saw boys guarding stacks of harvested wheat. They would fire their rifles as the flocks landed, hitting some of the birds. Since the birds in the flocks had fledged earlier in the year, the survivors were not easily discouraged and would merely move to a nearby stack. All that running around proved more effective in tiring the boys than frightening the pigeons. A similar predicament faced eleven-year-old Moses Van Campen when he was serving as "scare pigeon" one Sunday morning in September 1768 while his parents were at church. (The family lived near the Delaware Water Gap in northwestern New Jersey.) A short while after they left, masses of pigeons settled on a newly planted wheat field. His joy in chasing off the pigeons soon ebbed, and his actions had not been particularly effective anyway, as the flocks merely repositioned themselves from one part of the field to the other. Then he remembered the six-foot-long "fowling piece" that hung on the kitchen wall. The gun had accompanied the family from the Netherlands and was considered a prized heirloom. It was always loaded, so Van Campen merely took the rifle and crept outside where he could rest it on the fence railing near the pigeons. "He put his face down along the stock just back of the lock, sighted along the barrel and pulled the trigger. There was success at one end of the weapon and disaster at the other." Twenty pigeons were killed, but the kick from the gun flung him backward, skinning his nose in the process. But the real disaster came later when his father, ignoring Moses's valor in trying to save the family's produce, gave him "a flogging for having ventured to lay desecrating hands on that treasured" firearm.[26]

Even from a purely anthropocentric view, the guardians against pigeon depredation sometimes went too far. Officials in Quebec twice issued orders protecting landowners from the destruction wrought by the pigeon hunters themselves. In 1710, pigeon hunting was banned on land "that had been planted with wheat, peas, and other grains"; in 1748 authorities expanded that prohibition to property held by one particular landowner in order to protect his woods and fields.[27]

Occasionally farmers resorted to chemical weapons. There was no social harm in luring the pigeons with wheat soaked in alcohol, as one farmer did in Wisconsin. As the birds became soused, he merely loaded them into bags. There were, however, potentially deadly consequences to consider when the agent was strychnine or other poisons highly toxic to people. The pigeons might well survive for several hours and cover many miles before expiring, thus tempting numerous people with potentially harmful victuals. One of the most accomplished female bird students of her day, Jane Hine, recalled the spring in the early 1850s when passenger pigeons invaded the southern Lake Erie shore from Erie to Cleveland: "They produced a panic among farmers. They swarmed in oat fields recently sown and took the seed from the ground. They came into barns for grain." Everyone dined on pigeon pie until they heard that the birds might be tainted—farmers near Erie were said to be poisoning them. Fortunately, the scare ended when the birds moved on a short while later. A decade of pestiferous pigeons in eastern Minnesota led some farmers to lace grain with strychnine to vanquish the feathered hordes. Warnings against eating the birds were issued, as they had been earlier to discourage the consumption of pigeons that might have partaken of gopher poison.[28]

Happily, dealing with the pigeons did not rely exclusively on poison, guns, and clanging bells. It led to an important innovation in agriculture. Daniel Van Brunt, working in Horicon, Wisconsin, developed the first underground seeder in 1860, specifically as a way to thwart the pigeons.[29]

Passenger pigeons competed with farmers in yet one other way. Given the omnivorous feeding habits of swine, farmers allowed their animals to rummage freely through whatever habitat made up the neighborhood. During mast years, the pigs could fatten up on the nuts, obviating the need to provide grain. (In the absence of mast, another and less expensive alternative would be to bring skinny hogs to market at a lower price.) But an influx of millions of pigeons left little in the way of surplus for the semidomesticated pigs. In few places was this possibility of greater concern than east Texas, where sizable numbers of people lived at marginal levels during the best of times. That concern grew as the pig population in the state swelled from seven hundred thousand to almost two million over the three decades ending in 1880.

Whenever a big mast year began to draw pigeons, the newspapers of the region started voicing fears the birds would usurp the food, leaving none for the pigs. From Jasper County in 1875: "The wild pigeons are robbing the hogs of the mast." From Nacogdoches in 1881: "We think there is a sufficient crop of mast in some localities to fatten pork, if not destroyed by the pigeons and squirrels." But no one was more agitated over the avian threat to porcine welfare than a newspaper editor from Leon County, about sixty miles east of Waco: "Droves of one or two hundred each can be seen flying around prospecting, and it will not be many more weeks before there will be millions of them sweeping through the forest, eating all the acorns and causing a wail of despair to ascend from the throats of our beautiful razorbacks."[30]

PASSENGER PIGEON AS PRODUCT: MARKET HUNTING

It probably did not take long after the first European settlements in North America gained their footholds before people began selling passenger pigeons to their neighbors. The author of one early account from 1633 "bought at *Boston*, a dozen of *Pidgeons* ready to pull'd and garbidged for three pence." Some of that which was sold was likely the surplus of what was killed initially for home consumption. But over the centuries as human populations increased and technical advances allowed access to nearly everywhere, a trade in passenger pigeons developed that was maintained by thousands of people, most of whom were not pros, but merely locals who opportunistically took advantage of an easily made buck. This new situation intensified the plundering of the pigeons and ensured their extinction.[31]

Many are the tales of rural people bringing their pigeons to the nearest town for sale. In the 1820s so many birds were brought to the market at Quebec that large quantities went unsold and were allowed to putrefy on the street, even though a dozen could be unloaded for as little as three pence. To eliminate the health risk, city officials enacted a law requiring proper disposal. A few decades later near Terre Haute, Indiana, Tacitus Hussey wrote of the thousands of pigeons sent to market that brought a penny apiece, even though they were already dressed: "I have in some neighborhoods, seen wash tubs filled with these dressed birds carted off to village markets and sold at a price which would not pay for the time taken, if time was worth anything at all!"[32]

Sullivan Cook moved to Cass County, Michigan, in 1854 and set about carving out a farm from the forests. Taking a break one morning, he was roused by one of his daughters, who ran into the house exclaiming, "Pa, come out and see the pigeons." A massive flight was under way, as he observed "flock after flock of the birds, one coming close upon the heels of another." But he did not watch for long. He grabbed his shotgun, powder, and ammunition and ran with his twelve-year-old to a high point where he blasted away for a half hour before he broke for breakfast. A tally of the morning's work revealed twenty-three dozen birds. He was low on ammo anyway so he drove to Three Rivers and sold the lot for sixty-five cents a dozen. Given the high price he received and his location, the birds might have been destined for such cities as South Bend or Kalamazoo, or even Chicago.[33]

Markets blossomed at different speeds at different places. There is no evidence, for instance, that pigeons taken in Texas were sold out of state. In Ontario, a market for pigeons did arise, but it never reached the proportions that it did in most of the U.S. pigeon range. In the states, pigeons sold by the dozen, hundreds, or barrel, while in Canada the measure was usually by the bird, the pair, or less commonly by the dozen.[34]

Much of Ontario remained relatively undeveloped as late as the end of the nineteenth century. People devoted most of their efforts toward agriculture or other infrastructural activities such as lumbering. As one resident from Manitoulin Island said of the pigeons, they "came at a time when there was something else to do than disturb game." But where cities arose, so did local markets, and with time birds killed in such Ontario counties as

Middlesex, Simcoe, York, Lincoln, and Welland supplied urban markets. In Middlesex County locals took bushels of birds throughout the summer and early fall from the pines where they roosted. Pigeons killed by gunfire brought five cents each, while those that were netted or poled yielded an extra cent apiece. The hunters collected the carcasses and prepared them: "We usually hung the birds . . . overnight, to cool off, and packed them in layers of straw in apple-barrels." The product was then packed on boats to Buffalo. Detroit was closer, but during years when the species was thick in Michigan, the price was probably better in New York. Birds originating in other counties were stuffed in barrels and shipped to Toronto or Montreal.[35]

When the pigeons still appeared in large numbers in New England, local markets were in the big cities. Luther Adams was a "farmer and horticulturalist" who lived in Townsend, Massachusetts, just fifty miles from Boston. It is wooded and hilly country, and he conducted most of his netting on a high promontory. In 1847, he took 5,028 birds, and in 1848 almost 2,000. There is no record of what he received for them, but he was conscientious about listing his cost for bait and other expenses. In 1847, he bought nine bushels of buckwheat for $4.50, wheat for $3.50, labor for $1.00, use of netting sites for $3.50, and other expenses for $0.75.[36]

Regional markets were already in full operation by 1851. In that year, at least four different companies trading in pigeons shared the Plattsburg, New York, nesting. A careful count indicated that one million birds had entered the nearby town of Beekmantown over two days. The networks were already in place, for "the news of this congregation . . . soon reached the ears of the old pigeon catchers in different parts of the country." A firm from Massachusetts, the Harris Company, arrived first and began baiting, eventually reaching the level of four to six bushels of grain per day. (Over seven hundred bushels of corn and buckwheat would be used by all the parties during the nesting.) Then the other companies showed up, as did freelancers. The number of birds they shot or netted broke previous records: "It would be almost impossible to give an accurate account of the whole number taken; but four companies engaged in catching and purchasing, the writer knows, forwarded to the different markets not less than *one hundred and fifty thousand dozen*" (emphasis in the original). Live birds brought from thirty-one to fifty-six cents a dozen. Fifteen to twenty-five people were paid five cents a dozen to dress the birds.[37]

Efforts to create national markets for game started before rail service became available. Chicago, for example, received vast quantities of dead birds collected from nearby grasslands and marshes. But the dealers had no way to transport it east, as they had no ice to keep the product fresh during the summer when shipping on the lakes was an option. In November 1850 merchant Robert Saunders tried an experiment. A four-horse prairie schooner hauling 1,650 grouse (sharp-tailed and/or ruffed) and 3,500 prairie chickens in boxes left his warehouse at the South Water Street market and headed to Buffalo. The weather proved favorable, as temperatures were cool enough to preserve the carcasses. Much of the cargo sold in Buffalo, with the remainder being shipped to Albany and New York City. The driver made the lonesome trip back having earned for Saunders an acceptable profit, but presumably the risks were deemed too high to repeat, and the rails reached Chicago soon thereafter.[38]

National markets became established as the tentacles of the railroads penetrated more and more deeply into pigeon range. The first rail service in the country commenced on Christmas Day 1830 when the engine Best Friend of Charleston towed several train cars along six miles of tracks out of South Carolina's capital. (The engine exploded the following year.) Within the next three decades, the nation's length of track would exceed thirty thousand miles. During the 1850s, rail expansion linked New York City with the Great Lakes, Philadelphia with Pittsburgh, and Baltimore with the Ohio River at Wheeling. The Michigan Central and the Michigan Southern each made it to Chicago in 1852. Connections from Chicago to Galena and East St. Louis, both on the Mississippi River, were completed over the next three years. In Wisconsin, towns that would become famous for the vast pigeon cities nearby, such as Kilbourn City (now Wisconsin Dells) and Sparta, received rail service in 1857 and 1858, respectively. As early as 1842, three thousand pigeons from Michigan were delivered to a station, perhaps by boat, and then rode the rails to Boston. The extension of the Erie Railroad from New York City to southwestern New York opened up the pigeon trade from that area of frequent nestings, while in 1882 thirty-seven miles of new track allowed the easy transport of birds from Potter County, Pennsylvania.[39]

Birds from the most remote reaches of pigeon range could be transported to rail lines. The farther it was to go, the less profit the dealer made, but it would still potentially be worthwhile. The expanding coverage of telegraph wires made pigeons even more vulnerable, since a large gathering of birds not only would draw the attention of locals, but would soon bring people from far and wide. The agents who staffed the rail stations made it their business to spread the word. (As a small measure of justice, perhaps, the sheer mass of pigeon flocks sometimes threatened or actually brought down telegraph wires as the birds flew into them or attempted to perch.)[40]

The predicament that faced Stephen Sickles of Smethport, Pennsylvania, in spring 1842 illustrates the rise of markets in one area. The pigeons arrived in their throngs, and Sickles waited anxiously with his net ready to go at them. But there was no market then, so he had no way to sell what he caught. Instead, neighbors paid him $2 a day to catch the pigeons for them. Thirteen years after that, local markets were flourishing, but it still did not occur to people to transport to and sell their pigeons in the big cities. By 1880, however, things were very different in that part of Pennsylvania:

> By this time netting and shooting pigeons to be sold in the city markets had become a well organized business. Those engaged in the business were supplied with accurate information as to the locality where the birds might be found at any given time, with an estimate of their number and directions as to the most direct route by rail to a point nearest to the nesting place. This accounts for the great slaughter of the pigeons that took place during their nesting in the vicinity of Dingman Run.[41]

Two of the best-established pigeon dealers were the brothers Joseph and Isaac Allen of Manchester, Michigan. The growth of their business provides another example of how pigeon trading expanded. They were just boys when the family moved from New York to Adrian, Michigan, in 1854. Flocks of pigeons migrated through in steady numbers for close to thirty years. Their early hunting activities, however, were more for sport than commerce, and they liked nothing better for breakfast than "a nice broiled pigeon." They recognized that the price for pigeons would be highest in New York City, but their father didn't trust the express companies. On one particularly heavy flight day, their father had bagged six hundred pigeons by ten A.M. One of the sons was instructed to take the birds into Adrian and sell them for a dime a dozen. Being of sounder entrepreneurial timber than his dad, the son began offering them for twenty cents a dozen, and then a quarter, until he ran out. Later in the day, having replenished his supply, Dad disposed of the birds for the original dime he suggested. In New York City, the same twelve pigeons would have brought $2. The boys were tired of a measly dime or even a quarter, so during the next year's flight (mid to late 1850s) they resolved to ship them to New York. Their father objected, "It is foolish for you to send them, as they will never be heard from." But in just four days, they received their compensation: seventy cents a dozen. This was the lowest amount they were ever to get for pigeons, but this bunch was even more significant in being the first pigeons ever shipped from Michigan to New York City. Though small reward for a pack of pigeons, it was a giant leap toward establishing the Allen brothers as traders of national scope.[42]

In 1861 few people made their living chasing passenger pigeons. A partial list of names and where they lived indicate that most claimed New York State as home, with others from Columbus; Hooksett, New Hampshire; and Camden, New Jersey. But over the next ten to twenty years the number would grow substantially, although there was never agreement as to how many there were at one time. The highest figure offered was from H. B. Roney, who opposed the slaughter. In 1878 he said that five thousand men were professionals. An estimate made four years earlier had claimed six hundred, while another in 1880 gave the figure as twelve hundred. The Allen brothers said that between one and two hundred practiced the trade full-time: "The pigeon business was very profitable for men who were used to it . . . When the pigeons changed their location, the pigeoners would follow them, sometimes going over a thousand miles."[43]

Pigeoners doggedly pursued a large roost that settled into the forests of Shannon, Oregon, and Howell Counties in southern Missouri in early 1879. The birds drifted north between eight and twelve miles a day with the men in hot pursuit. When the birds began their rest for the night, hunters would work in close coordination with each other. They would scatter through the tract, waiting for a prearranged signal before starting to fire into the trees. The carcasses were collected in the morning and delivered to Piedmont, the closest station on the Iron Mountain railroad: "From here are shipped every day from seven hundred to one thousand dozen pigeons, bringing into the county from six to eight hundred dollars, net cash per diem. The birds are sent to Boston and New York, where they sell at $1.30 and $1.60 per dozen."[44]

The birds did not always make it easy for the professional pigeoners, however. Some places proved difficult to operate in because of terrain, distance from transportation nodes, or other factors. Pennsylvania, for example, drew lots of pigeoners and produced huge numbers of dead birds, but with its rough topography even the most experienced hands caught far less than they would have expected. A nesting in Emmet County, Michigan, occurred in dense cedar swamps reached by almost unusable tracks. H. T. Phillips, whose Detroit commission house handled its first live pigeons in 1864, tells how he transported live pigeons that had nested on the Black River near Cheboygan, Michigan, in 1869. He placed the birds in the crates that he had just constructed out of split cedar. The crates were then loaded on two canoes that had been tied together. Phillips and his birds shoved off on a six-mile float downstream to a dam, at which point he had to transfer the rig to the other side. After twelve more miles they reached the steamer at Mackinaw City that would take them to yet another boat that would complete the final leg of their trip to Detroit. At least his birds arrived at their destination. In another instance, Phillips had to discard three thousand pigeons "because the railroad did not have a car ready on the date promised."[45]

James Bennett began his career as a pigeoner in 1867. He and his uncle lived in Lycoming County, Pennsylvania, where the pigeon flights were an expected annual occurrence: preparations for catching the birds were as ordinary as the sowing of crops. But that year, he decided to follow the birds and would continue to do so for fifteen more. On September 15, 1877, he headed southwest, prepared to go as far as the Indian Territories (Oklahoma) to find the roosting pigeons.[46]

Bennett began searching in earnest in Verona, Missouri, southwest of Springfield. Having gained the use of a wagon and two horses, he methodically explored valley after valley until he arrived in Cherokee, near the border of Arkansas and the Indian Territory. Word was that the birds were in a tract of blackjack oaks fifteen miles by forty miles, just two days to the west. The information proved true, for on the night of the second day he came upon the pigeons, "craking and flying in such numbers for about a mile before I reached the roost."

The sky was clear and the moon reflected with bright intensity. Not being able to wait, he fired his shotgun into the trees and brought down forty-one birds. Nine Indians were there, too, collecting pigeons for various dealers. In three nights of shooting, they bagged and sold 3,630 pigeons. Bennett stayed all the way until February. Then it was back to Pennsylvania for the spring flight: "That season there was carload after carload shipped from Kane and Sheffield to the northern market."

A few years later he finally called it quits, "leaving the forests of Potter County, on the Coudersport Pike, May 29, 1882." There just weren't enough birds to make it worth his while anymore, although he did get a card in 1888 announcing that "scout" pigeons had arrived near Sheffield. But the scouts either did not like what they saw, represented all there were, or never made it back to their waiting comrades, for that was the last Bennett heard of them.

Another professional, W. C. Waterman, held out a little longer. He saved the last sales receipt he ever received for passenger pigeons. On April 25, 1884, the firm Bond and Pearch of 163 South Water Street in Chicago credited his account in the amount of $90.46 for two barrels of pigeons, 523

"January 5, 1878, The Christmas Season/ Game stand, Fulton Market, NY."
The artist is A. B. Frost and this first appeared in Harper's Weekly, *date of issue unknown.*
Courtesy of the New York Public Library

birds, procured near Madison, Wisconsin. The total reflected their subtraction of $7.60 for freight and commission.[47]

Modern commentators emphasize the importance of the national markets in driving the pigeons to extinction. Some have stressed that once the pigeon population went below a certain point the professionals, such as Bennett, had to throw in the towel. Thereafter, the argument goes, the birds were not heavily exploited. This is not true, however. Amateurs took a huge toll on birds and kept it up so long as they encountered pigeons of any size population. They did not necessarily chase the pigeons far, if at all, for it is likely there were people to catch and people to buy within a day's ride of wherever the pigeons put down. Edward Howe Forbush wrote, "*From soon after the first occupancy of New England by the whites until about the year 1895, the netting of the Passenger Pigeon in North America never ceased.* Thousands of nets were spread all along the Atlantic seaboards" (italics his). The birds, he correctly states, were caught wherever they appeared, both before there were regional and national markets, and within a few years of their disappearance as a wild species.[48]

Except at the big nestings, these folks who supplemented their livelihoods by the temporary, if not fortuitous, presence of the pigeons neither attracted much attention nor left much of a record. There was William Armstrong, who related that a majority of his neighbors near Blairstown, New Jersey, netted during pigeon season. Lucy Bennett was another who left information on the occasional killing of pigeons for economic gain. She and her husband, John, lived in Decatur, Michigan. She kept a diary, and from February 12, 1873, to April 9, 1875, she made at least twenty entries related to John's pigeon hunting. Early in the season, usually February, he worked on his net. In 1873 he knitted one that was sixteen by thirty-two feet. Many days he came back empty-handed, but on a few he scored large hauls. On April 7 of that year he caught forty dozen. These were plucked the following day and shipped out on April 9, the first the Bennetts seem to have sold in such a manner. For March 31, 1875, she notes, "We arose at half past 3 o'clock, got breakfast, then John and Peter took the horses and went pigeon hunting. Hope they will catch some . . . John caught several dozen pigeons today." Although John captured a few the following day, the wind blew too hard for his efforts to be very fruitful. The final entry is April 9, 1875: "John came home just as I did. He caught 27½ Doz. Pigeons today . . . I think all the money they will bring is dearly earned."[49]

Jennifer Price creates a compelling vignette of how a pigeon from Sparta, Wisconsin, would get to a restaurant table in New York City in 1871: "One might well begin with an eight-mile wagon ride one morning from the northwest edge of the nesting to the Sparta rail depot. The squab is packed in a barrel of ice and shipped on the 3:00P.M. Milwaukee–St. Paul express train to Milwaukee and then south to Chicago, where the barrel is transferred to the Chicago, Burlington & MO express train to New York City. A driver for the American Merchant's Union Express picks up the barrel at Grand Central Station and delivers it to a game dealer, who has purchased the pigeons from a Chicago dealer on commission." New York City had two game markets at the time, with Fulton's being the choice of the higher-end restaurants, including Delmonico's. A representative of the restaurant, often Lorenzo Delmonico himself, would visit early every morning to select the fare that would be offered that day. On that particular day, his wagon would carry pigeons, as it made its way "from Fulton and South Streets up Broadway to Fourteenth Street and is unloaded in the restaurant kitchen on Fifth Avenue."[50]

What went into that sojourn goes to the heart of the nineteenth-century game business. A unique perspective into that business is provided by H. Clay Merritt, who jokingly claimed he was born with a gun in his hand. Merritt was unusual for men of his profession in having had a first-rate education. He attended a prep school and then spent four years at Williams College, where he received his degree. Throughout it all he loved to hunt, and after college he moved to Henry, Illinois, to make a living out of his great passion. Passenger pigeons made up only a small part of his trade, which was dominated by ducks, shorebirds, bobwhite, and various grouse. The key to his success was not his ability to procure birds, but to sell them when the price was highest. During, say, woodcock season, the price was low, for everyone had the plump little fellows for sale. Merritt would hold his stock for varying lengths of time up to and beyond a year until he was comfortable with the price. To hold them that long, he needed ways to store and ship the birds that kept them edible.[51]

Keeping birds reasonably fresh was one of the great challenges of the game industry. In Minnesota, for instance, the period from 1870 to 1900 saw many warm autumns that "played havoc with the shippers . . . for the game often spoiled before it reached St. Paul." Merritt tried numerous techniques to freeze product. He would cool birds on ice and then pack them into boxes so tightly they retained their temperature for several days at least. When the birds were obtained in winter, he could freeze them at ambient temperature and hold them until prices rose. In 1870 he went out to Sandusky,

H. Clay Merritt. Photo from *Shadow of a Gun* by H. Clay Merritt

Ohio, to inspect a new model of freezing room designed for butchers. He bought one and then two others. Eventually, he hired a carpenter to construct one of his own design that was built belowground: "Though they were inferior to those that came later, we could and did freeze birds fairly well in summer."[52]

Shipping birds as cold as possible enabled them to be sent dry. This was an important goal of Merritt's, for any ice accompanying the packages cut into profits by raising shipping costs. In fact, he was often warned that unless the shipment was iced, it would be lost. But given that he would ship up to twenty thousand birds at a time, it was a risk worth taking. Merritt was proud that he only lost one entire batch. More commonly, the birds became moldy and even began to smell. With just a little scraping, though, the product was once again salable, albeit at a reduced price if the birds had ripened too much. One fall Merritt received an order for snipe, which he had many hundreds of from the previous spring. This request was from a good customer, so he went to the freezing bins where the birds had been stored and discovered they were all in bad condition. Still, to satisfy his buyer, he sent a box of four hundred and threw away the rest, in no worse shape. He kicked himself for the latter action when he received a check for over $100.[53] The challenge of sending meat long distances and having it arrive wholesome was finally resolved in 1878 when Swift and Company, a Chicago-based meatpacking firm, introduced the first successful refrigerator car.

Merritt dabbled in passenger pigeons, but they were apparently local enough by the mid-1870s that he no longer encountered them in his ordinary hunting activities in central Illinois and points west. He, therefore, bought them from W. W. Judy, "the ruling game dealer in St. Louis." Pigeons were his specialty. Merritt purchased several thousand birds for seventy-five cents a dozen. The next time around, in 1881, Judy charged only fifty cents a dozen, and Merritt procured "a good carload," which arrived in batches of fifty to seventy-five. These were frozen and kept until the next spring, when they were sold for two to three times as much. His final purchase of pigeons occurred with Judy's death. Merritt bought "stall-fed" birds—pigeons kept and fattened in cages before being killed. These sold for as high as $3 a dozen. He held some for as long as three years: "The last barrel we marketed in Boston at full price . . . We believe that these were the last stall fed birds that were ever marketed."[54] If only he knew that at the time, for he could have auctioned them off to some baron who would have paid handsomely for the privilege of eating the last of a foodstuff, if not the last of a species.

Live birds for trapshooting brought the highest prices. This trade depended more on national markets than perhaps even the business in dead birds, for the big shooting meets varied in time, place, and sponsors. The market for live birds was surely smaller than for dead ones, but the waste was much greater, for although match organizers would utilize pigeons in awful condition, the birds had to be at very least alive. Of course even a blob of fermenting meat and feathers could be propelled through the air by a plunge trap, and if the shooter was quick, maybe no one would notice. Perhaps it was the heat, but some of the most horrific reports of waste came from Missouri: of sixty thousand birds secured for a trap meet, two thirds died before they reached their destination, while on another occasion only thirty-three hundred out of twenty thousand wound up as targets.[55]

Thomas Stagg focused his business on the trap trade. He maintained his home and operations on forty acres between Fullerton and Diversey on the western edge of Chicago. He had removed the external sidings of a barn and replaced them with a lattice of narrow wooden strips, thereby converting the structure into a giant cage that could hold over five thousand passenger pigeons. The selling price for his product was on average $1.25 per dozen. Stagg and an assistant would travel frequently to nesting grounds in Michigan and

Wisconsin to procure birds. They entered the pigeon cities at night and grabbed low-roosting birds by hand, depositing them into bags. From the bags, the birds would be placed into crates and then shipped to Stagg's place. Upon reaching their destination, the birds were often parched, and "many drank themselves to death, or were killed in the mad scramble for water." Sometimes the birds did not arrive at their final destination in the best of condition, as was the case with a load of thirty-five hundred that Stagg sent to New York. They arrived with feathers and skin missing from their heads as a consequence of their having rubbed against the crates.[56]

One of the largest dealers in the country, who marketed both live and dead birds, was also in Chicago. Bond and Ellsworth on South Water Street stocked barrels of dead pigeons on the first floor of their warehouse, which had walls adorned with elk antlers. A visitor was surprised that he saw no live birds, but upon being escorted to the upper three floors, he saw where they all were. Each floor held twelve cages of twelve feet square. The cages were equipped with perches to allow fuller utilization of the available space. Ordinarily, fifteen thousand to twenty thousand pigeons could be accommodated, but during emergencies additional birds could be squeezed in. Mortality among newly arrived birds amounted to thirty-five to fifty-five pigeons in every cage over the first two days.[57]

How many birds the pigeon industry destroyed is impossible to know. Edward Howe Forbush tried to get information on the netting of the pigeons by the major firms, but by the time he tried, the firms had already dissolved. He sought the help of Otto Widdman, Missouri's premier ornithologist of his day, to contact W. W. Judy and Company of St. Louis. He found that all but one of the partners had been dead for a while, and the survivor offered nothing except the belief that all the passenger pigeons had flown to Australia. Old ideas may be harder to kill off than hundreds of millions of birds.[58]

CHAPTER 5

Means of Destruction

History suggests that few things stimulate human ingenuity more than the challenge of killing. This is most evident when the intended targets are other human beings, for no other organism poses anywhere near the same severity of threat. But as a species, we are no slackers even when the adversary is an eighteen-inch-long bird. Safe only when they rose high enough to exceed the range of weaponry, the passenger pigeons otherwise lived a gauntlet whereby they became targets of an arsenal that employed an amazing array of instruments.

In rare instances the birds were poisoned. Asphyxiation was tried by burning sulfur underneath nests. All one needed was five or six fireproof containers, two ounces of sulfur, a few sacks, and a torch. Arrive at the roost after dark, distribute the sulfur among the containers, and then ignite the contents. Stand to the side to avoid the fumes and wait for the birds to drop from the trees: "This hunting is easy. Women can take part with pleasure since there is neither fatigue nor danger of being wounded."[1]

Fire was also used in at least two ways. The first was described by William Bartram, one of the country's first great naturalists. He arrived at the villa of a friend near Savannah, Georgia, just after dark. Soon, servants showed up with "horse loads" of pigeons collected over a short period from a nearby swamp. They had entered the roost with blazing torches, the light of which so blinded and confused the birds, many dropped to the ground helpless. The servants then easily gathered the prey and put them into sacks.[2]

The second way was simpler and more effective, for the fire did not merely leave the birds dazed, but dead. Employing the "grand mode of taking them," a roosting site in Tennessee was set ablaze, incinerating swarms of birds as they wheeled furiously in their confused attempt to escape. From heaps two feet deep, scorched corpses were then collected the next day for personal use or sale. Texas saw one of its few large pigeon invasions in the fall of 1872. Despite ongoing exploitation, the birds stayed into the spring, when they attempted to nest in thick stands of Ashe juniper. (Birds that fed on juniper cones were said to taste like turpentine.) But farmers, leery that their crops would be endangered by the enormous number of the feathered immigrants, burned thousands of acres of woods to rid themselves of the menace.[3]

Low-flying pigeons could be downed by just about any object at hand. An anonymous Jesuit wrote at length about pigeons along the St. Lawrence River in the mid-1600s: "They passed continually in flocks so dense, and so near the ground, that sometimes, they were struck down by oars." Clubs and stones killed pigeons as they flew low over St. Paul, Minnesota, in June 1864. The strangest weapon used to dispatch the birds was employed by settlers in Orillia, Ontario. While harvesting their potatoes, farmers took advantage of nearby birds by flinging tubers at them. It is good to read that they lost more potatoes than they gained pigeons. But to have most of a stew fall from the sky in one lump must have been convenient.[4]

At Racine, Wisconsin, during September, the pigeons would pour south in dense flocks along the Lake Michigan shoreline until they reached the Root River and followed its course inland. Often the birds would fly just a few feet above the ground and not higher than forty feet. Enterprising residents took advantage of the situation in several ways. Slightly more sophisticated than sticks, garden rakes and pitchforks enabled many to harvest their fall crop of pigeons. Two brothers preferred fishing for their birds. They stretched a hundred-foot-long seine and secured it to the tallest branches they could reach. The mesh proved invisible to the birds as they sped along, and large quantities became entrapped, while others fell to the ground stunned. The net would be released and fall to the ground heavy with entangled pigeons. More effective still was the "pigeon killer," as the operators called it, erected on an elevated bank of the river. The simple design consisted of a "long hickory pole in the ground" with cords stretching from it in opposite directions. When the birds "passed over the bluff the boys would vibrate the pole rapidly by pulling the cords alternately, the top of the pole knocking hundreds of them to the earth." The kids would work in teams, some manipulating the pole while others gathered the bounty.[5]

A similar pigeon killer was employed during the early 1800s to take birds that roosted at the celebrated Bloody Run or Pigeon Roost swamp in the Buckeye Lake area east of Columbus, Ohio. As the pigeons piled into the swamp at dusk, a long pole placed at the edge was waved around to bat all the birds desired out of the air. Over the decades, though, the effectiveness of this method waned as the birds became more wary: they no longer headed directly into the swamp but ascended beyond the range of poles and guns and dropped into the roost in a more abrupt trajectory.[6]

Bows and arrows rarely figured in pigeon hunting by white people, but if used, it was usually as a cost-saving measure over guns. One participant from the Timiskaming area of Ontario said guns were preferred only when it seemed certain that a single shot would bag five or more pigeons. Otherwise, the cost of powder and shot became prohibitive. Boys too young to have guns or the money to afford ammunition composed another group of bow-and-arrow aficionados. C. A. Fleming of Grey County, Ontario, wrote a long memoir of his experiences in the 1860s. He and his friends chose eighteen-inch-long shafts of cedar and hammered a nail into one end of each. The protruding nail head would then be ground to a fine point. Rock elm provided the wood of choice for the bow. The shooters armed themselves with twenty-five or thirty arrows, knowing that no more than one out of twelve shots was apt to puncture a pigeon.[7]

The Hussey boys grew up near Terre Haute, Indiana, during the Civil War. They, too, relied on bows and arrows for their pigeon sport: "When the great flocks of wild pigeons flew across the country so thick that you could not see the sky, we would send our arrows among them, and if it did not hit one going up, it would surely hit one coming down; and we would gather up the dead and wounded with that heroic feeling of boys who have been out and killed something."[8]

Where the birds roosted or nested, killing as many as one wanted was usually a cinch. Some hunters used their bare hands, while others bludgeoned their prey with clubs. Mark Twain recalled the roost near his childhood home in Hannibal, Missouri, and how all the pigeoners relied on clubs. Adults and squabs fell from their nests as they were slammed by poles. More profitable still, men cut trees loaded with nests that would in turn knock down other trees equally endowed. It was a simple matter to gather the fallen fruit.[9]

A TEMPTATION TOO STRONG FOR HUMAN VIRTUE TO WITHSTAND: SHOOTING PIGEONS

> In their passage the People of New York and Philadelphia shoot many of them as they fly, from their Balconies and Tops of Houses.
>
> —MARK CATESBY, 1731

What a shame that passenger pigeons became extinct. Future generations would be denied the near euphoria that apparently accompanied raising a gun toward a flock of pigeons and firing. Anne Grant said of the spring and fall flights that ascended the Hudson River and passed over Albany in the first years of the nineteenth century, "This migration . . . occasioned . . . a total relaxation from all employments, and a kind of drunken gayety, though it was rather slaughter than sport."[10]

A couple of decades later in York, Ontario (now Toronto), the arrival of the pigeons triggered another outburst of orgiastic firing. For several days, the city took on the character of a war zone, with the nonstop cacophony of discharging firearms resounding everywhere. Police attempted to enforce the ordinance banning the use of guns within the city, but it proved impossible given the sheer numbers of transgressors, including those of such high status as city council members, crown lawyers, and even the county sheriff. The forces of law and order capitulated: "It was found that pigeons, flying within easy shot, were a temptation too strong for human virtue to withstand."[11]

Urban pigeon shooters in Quebec City became so vexatious that municipal authorities appeared resolute in their 1727 enactment of a ban on such activities. A translation of the ordinance, written in the formal, breathless style typical of legal prose of that period, is too entertaining not to quote at length (to make it easier to read I have added a few words and a little punctuation):

> On account of the complaints which we receive daily from many people who spend their days here in various parts of the city of Quebec and as much as they trust in the security which accords with being in a city closed and policed they have nevertheless received blows from shot which have reached them. Others have gone into their yards and have wounded fowls and ducks which happens only because whenever there is a flight of pigeons and because of the eagerness to have them without taking the trouble

> of going out [of town] and going to those places where hunting is permitted, everyone takes the liberty of shooting thoughtlessly from his windows, the threshold of his door, the middle of the streets, [and] from their yards and gardens. [They do so] without thinking not only of the danger in which they place the passerby, old people, and the children who cannot take shelter sufficiently quickly from the danger to which they are exposed by indiscreet and clumsy people of whom the greater part know nothing about the handling of guns but more to the danger which they run for themselves in setting fire to their own homes and to other houses of the city as has happened several times from the wads of the fire-arms which have fallen all lighted upon the roofs of the houses. *We now make* express prohibitions to the day laborers and apprentices against leaving their work on workdays to go shooting at all either within or without the city under penalty of a fine of fifty livres for those who are in a condition to pay and a fine of ten livres and fifteen days in prison for the others."[12]

In both St. Louis and St. Paul the appearance of the pigeons brought out the shooters even though it was in violation of local law. A minister in Chicago during the 1840s complained that he could not write his sermon because the constant firing by his pigeon-seeking neighbors proved too much a distraction. While the college students of today have a plethora of things to keep them from their studies, they have been spared the allurement that tempted Samuel Cabot, who would later become a prominent physician and businessman. As he walked across the Harvard College campus one spring day in the 1830s on his way to a recitation, he was entranced by flocks of pigeons streaming across the sky. No doubt fidgeting throughout the duration of his class, when it eventually ended, he headed straight to his room, where he picked up his gun and joined shooters on a nearby ridge. In a brief time he shot eighteen birds.[13]

One cannot possibly evaluate the authenticity of claims for record shots. The big blunderbusses of those early times were very different from the guns of today—some of those old guns were so large they were fired from swivels. Even with that in mind, the claim made by a friend of Cotton Mather's that he killed 384 pigeons in one shot strains credulity, although the nature of the gun is not known. A more modest figure, and more plausible, is that reported from the St. Lawrence River territory. During the heavy flight year of 1662 a hunter shot 132 birds in one blast. Another Canadian much later told of killing 99 birds in one shot. He was asked why not 100, and he answered that he would not lie over one pigeon.[14]

Without question, the single discharge that killed more pigeons than any other occurred on the north shore of Lake Ontario sometime before 1846. The weapon was a cannon: "One of those prodigious flights came over Lake Ontario, in a direction for one of the garrisons, which being observed by the soldiers, a cannon was loaded with grape shot, and when the pigeons came within range, the contents were discharged amongst them, and made very great slaughter."[15]

But killing the birds was a cinch even to the majority without ready access to heavy artillery. Pehr Kalm in his travels noted the huge targets presented by massing pigeons and concluded that so "poor a marksman as to fail to make a hit is difficult to find." If one was in proximity to a roosting or a nesting, little effort was needed to shoot a lot of pigeons. One writer from New York provided instructions that pretty much came down to this: enter woods thick with pigeons, point gun muzzle up (that is, away from ground), blaze away, and, voilà, pigeons will fall at your feet and hopefully not on your head. Squabs were easy targets, too, although if the birds were too young and the shot too coarse, all that would remain would be gooey smithereens. But one team of young men working the nesting of 1860 in McKean County, Pennsylvania, managed, through trial and error, to get the technique down pat. Two of them carrying axes would pound a tree festooned with nests, and when the startled squabs extended their necks to peek out, the third would blow their heads off with his double-barreled rifle.[16]

Many hunters, though, sought to maximize the number of birds they could get per shot by converting the birds into better targets. During the 1770s, Canadian hunters would enter pigeon roosts during the day when the birds were feeding and install ladders on the sides of large pines. When the birds returned, they would take advantage and fill the new perches. After dark, the hunters would sneak back into the roost and begin shooting up the ladders, killing many more of the tightly packed birds than they otherwise could.[17]

Another account of nocturnal pigeon shooting comes from the early 1850s. After dinner and making sure their horses were secured, C. W. Webber and a friend entered an autumn roost that was over five miles in length in the Barrens of southern Kentucky. Though night, enough light seeped through the trees to provide this stunning description:

> Here we are among them! Look at that huge, low black mass—it looks like a great wall, several acres wide. One, two, three, fire! in

> platoon. I hear no sound—surely our guns missed fire; stunned and amazed, it seems a wild dream—that black, heavy-looking wall springs up like magic, and a tall wood is there—while, with a noise of wings, that made the earth tremble, lifting themselves into the dusky air—filling it confusedly as snow-flakes fill the dimmed moonlight of a winter's storm—the birds nearest us move off; but myriads take their places; and, while we rush in with lanterns, and with torches, to gather up the dead and wounded, the young wood is bowed again into our very faces; and, lifting our lights we can see the birds, clinging in the hundreds, to the limbs within our reach—their bright, black eyes dazzled by the glare, and they, uttering that soft, mellow cry, with a quick, incessant iteration."[18]

The early residents of Connecticut called the first crisp mornings of early fall "pigeon mornings," for those were the days when the flocks of pigeons would be expected as they migrated down the coast. In preparation for the birds, hunters climbed the low hills east of New Haven and secured long poles to the tops of the largest trees so they would jut out at a thirty-degree angle. On the forest floor, the gunmen constructed blinds where they could hide until the poles were filled to capacity with resting pigeons. During a good flight the withering fire dropped enough pigeons to fill a hay wagon before breakfast.[19]

But it wasn't always easy to bag a pigeon, for they often proved surprisingly resilient. C. A. Fleming advised that it was a waste of ammunition to fire at a flock coming at you. The shot was unlikely to pierce the thick breast feathers, and it would be difficult to see a bird fall as it would already be past you. Better to wait until the birds were heading away. Others disagreed, feeling the best chance for success was in firing at the head of the flock.[20]

An anonymous writer from Wisconsin discussed the hardiness of the bird in detail, based on the week he spent at a nesting where he and others "carried out a wholesale slaughter, which, I confess, partook of the nature of sport to the extent of making enormous bags." He found the pigeons to be "peculiarly tenacious": despite their small size, shot finer than a No. 6 would unlikely prove fatal. Even with a large-size shot, the birds would probably fly a few hundred feet before landing in a tree and falling over dead. A close examination of the carcasses he cleaned revealed a host of wounds from "previous assaults": "Broken and disjointed legs; bills that have been shot half away and grown curiously out again; missing toes or even a whole leg; and even healed up breast wounds."[21]

A. W. Schorger concluded that of all the techniques used to kill passenger pigeons, shooting claimed more birds than any other. But gunfire also prevented the final score from being *Homo sapiens* five billion, passenger pigeons one. The earliest I know of such fatalities occurred near Mount Holly, Pennsylvania, in March 1740. "A young lad who had been shooting pigeons, hanging a parcel of them over the Barrel of his Gun, flipt down to his Trigger . . . and discharged the Piece against his Breast, and killed him on the spot."[22]

A perusal of Wisconsin and Michigan newspapers reveals additional human casualties associated with the destruction of the pigeons. Only one does not involve firearms, and that came from an interview Schorger conducted with an elderly gentleman in 1936, who as a boy helped collect squabs at a nesting near Kilbourn City, Wisconsin. Groups of Indians used to work the nestings, including kids who climbed trees to reach squabs otherwise inaccessible. In this case, the branch broke and the youngster fell, breaking his back as he struck a log on the ground. He died almost instantly and was buried on the spot by grieving family members.[23]

But otherwise the injuries were mostly from errant shots. In 1844 near Prairie du Chien, Wisconsin, Samuel Gilbert's twelve-year-old son was part of a group running with guns cocked toward a flock of feeding pigeons when he was struck by the discharge of a friend's weapon. An unusual fatality was the man shot while hunting pigeons in 1853 near Waukesha, Wisconsin. Frank Crandall out of the Baraboo area injured himself when his gun went off accidentally.

Many woundings occurred in central Wisconsin during the huge nesting of 1871. The *Burlington Republican* reported in its March 18 edition that Gregory shot at a flock of pigeons with one barrel but hit Hanrahan with the other. The March 29 issue of the *Janesville Gazette* highlighted two recent shootings: a youngster in Lafayette County inadvertently shot his friend, while in Wautoma, Frank Clay was the unintended victim. Later that year near Cedar Run, Chas Harting was also wounded by a hunting companion. The one Michigan report, from the *Buchanan Record* of April 27, 1871, tells of a young man out pigeon hunting who pointed his gun at his younger brother's face, and it accidentally discharged. Fortunately Dr. Bell removed the shot and dressed the wound, preventing it from becoming serious.

Almost halfway between Racine and Kenosha in southeastern Wisconsin is the town of Somers. There in the fall of 1871 William Somers asked two young men to desist from shooting at pigeons on his land and was shot for his troubles. In late September of the following year, sixteen-year-old Frank Babcock suffered mortal wounds while hunting pigeons when he accidentally shot himself, according to the *Platteville Witness*. About the same time, a man was deemed to have committed suicide near Milwaukee. Next to him "lay a string of nine wild pigeons," although the connection, if any, between the pigeons and the act was unclear. From Dodgeville, Wisconsin, in spring of 1873 came news that the fourteen-year-old son of Samuel Klegg was shot by his brother as they hunted low-flying pigeons. Severity of the injury was not stated. I find it surprising that given the numbers of armed pigeon pursuers over the centuries, some not always sober, there were so few human casualties.

Hunting pigeons did exact one other casualty—the truth. Collectors of tall tales found some doozies. A hunter coming upon a row of pigeons perched on a low branch could carefully aim just so to split the limb, which when it retracted held the birds fast by their feet. All he had to do then was ascend the tree and cut the limb. Another time, a bunch of birds were feeding on wheat left by a thresher. A hunter carrying an 8-gauge shotgun crept up to the pigeons, and when they flushed, he fired. But to his amazement, not a bird fell. He examined the ground more closely and found numerous pigeon feet. He had fired too low![24]

A minister from Christian County, Missouri, said it was a waste of ammunition to fire into a large pigeon flock: the birds were so densely packed as they flew, a dead one could not fall. Another hunter working a pigeon roost in Howell County, Missouri, was after a bobcat, one of the predators that often appeared at pigeon gatherings. He tied his horse to a branch weighed down by a huge pigeon flock, then began his search for the larger game. Following a track, the hunter spotted the cat and managed to get off a shot. With the report of his rifle, the pigeons took to the air, and the branch, freed of their weight, snapped upright, leaving the unfortunate horse hanging by its reins. The luckless animal dangled there until the hunter returned with an ax and felled the tree.[25]

TRAPPINGS: NETS, BAIT, AND STOOL PIGEONS

> No one other factor contributed more to extinction of the species than did organized netting.
>
> —DUANE YOUNG, 1953

Traps proved to be both more efficient and less expensive than guns in capturing pigeons, and trapping was the only way to satisfy the live-pigeon market. (By the 1870s, though, some netters were finding birds too spooked to approach bait.) Almost all of the traps involved spring-nets, but a few were of other design. Newly fledged squabs could be lured into pens by live decoys, then their exit would be blocked with a net. Boxes of various sizes and even troughs used to collect maple syrup were raised at one end by props tied to long ropes. Grain was placed under the container, and when enough birds began feeding, the rope was released and the box enclosed the birds. A variant of this type proved successful in a wheat field near Winnipeg, Manitoba. Nets twenty feet long and fifteen wide were extended on frames, one side of which was kept up by an eight-foot-long pole connected to a cord that would be tugged at the right second. Would-be trappers placed dead trees and stuffed pigeons next to the net to lure the birds.[26]

From Ontario came two novel, albeit simple, approaches. In one, boys built small huts out of grain-bearing sheaths. When pigeons came to feed on the roof, the boys would reach up and grab them. A candidate for most horrible trap, although apparently not often used, was a platform about eight feet tall filled with sticking wax and enough food to attract the birds. The prize, though, for opportunistic netting has to go to the farmer in Massachusetts in the 1600s who told his friend Cotton Mather that he had caught two hundred dozen pigeons with two minutes' worth of effort: the birds flew into his barn and he merely closed the door![27]

A few enjoyed the sport of netting, but most pursued it for money. Some of these latter were professional, but many saw pigeon netting as a sideline, a way of augmenting the family larder. William Armstrong began compiling a list of netters in the vicinity of his hometown of Blairstown, New Jersey: "It soon resembled that of a list of voters at a polling place."[28] I think it has been overlooked that a great number of people killed passenger pigeons in their spare time and sold the surplus.

The netting operations were often intricate, involving captive pigeons, bait, and net traps. Where the birds nested, there were at least two variations. Early in the season when the pigeons were concentrated, nets were set up nearby to capture birds as they made their various foraging flights. This activity was known as "flight-catching," and most of the captives were killed. Later during nesting, as the birds dispersed from the nesting site, more time was necessary to attract a sufficient number to a given trap location. To compensate for the fewer numbers caught, most of these birds were kept alive for shooting contests, which brought a higher price per bird. Special pigeon baskets woven of hickory, oak, or other wood strips held live pigeons. One scribe called them a "portable prison house for feathered innocents." A narrow opening and neck enabled the hunter to easily stuff a bird headfirst, but stopped the birds from escaping. And wide potbellies prevented suffocation.[29]

Nets varied in size. Peter Yarnell and his brother used a tiny one of four

A pigeon basket of white oak made to hold and transport live passenger pigeons. Courtesy of School House Museum, Ridgewood Historical Society (New Jersey)

square feet. They still caught 21 pigeons at a time and 103 in a day, which exceeded their personal needs. But far more common were the industrial-size nets, which often ranged from twenty-five to forty feet long and ten to twenty feet wide. A net used in Ontario was twice that large. The best of the nets were made of linen, fashioned from hand-spun flax into meshes of one to two inches. On occasion the nets would be stained with butternut bark to make them less conspicuous. Two nets were sometimes used side by side and triggered so they snapped back toward each other making it harder for birds to escape.[30]

To get a sense of how these traps worked, picture a giant mousetrap with one end of the net attached to the bar. When it was released, the net would be unfurled over the birds attracted to the bed. Here is a description of a typical rig used in Pennsylvania: "One side of the net would be staked along the entire length to the ground and through the other side which was free, was run what was called the net string, which was fastened on each end to the spring poles by which the net was sprung. The spring poles were . . . doubled back to give force by which the net was spread and were a number of feet from the net. The net would be tucked carefully on the ground along the staked side and so arranged that when it was released it would fly out and spread itself over the ground or bed on which the pigeons . . . would alight." Sometimes the pigeoner dug a slight trench in which he could obscure the rolled-up net. In the same vicinity, two short stakes or "release rods" were

A bough house or hiding place for hunters, by the painter Arthur Tait, published by Currier and Ives in 1862. Courtesy of Garrie Landry

hammered into the earth. Each of these stakes had a shorter dowel driven through it that could be pulled by a cord held by the trapper to spring the net. As for the poles, trappers liked beech, cedar, hemlock, hickory, and other trees whose wood was both flexible and strong. The nets were often weighted at their edges to facilitate their unfurling and to make it harder for birds to escape; sometimes metal rings placed around the poles were connected to the ropes so when the net rested at the end of its trajectory, the ring would slide down to hold the birds more securely.[31]

Since the trap was sprung by a human hand, the bodies attached to these hands had to be hidden in a blind. This was usually called a "bough house" or "booth." It tended to be a small simple affair situated close to one of the spring poles with a frame of branches, often cedars and other evergreens when they were available. Then smaller branches, bearing a thick growth of leaves, would fill in the cracks, making the inhabitants invisible. Leading to it were a number of cords enabling the netters to manipulate the live pigeons that were so integral to the effort, as well as the trip rope.

All that work would be for naught unless the nets landed over lots of pigeons. The chances that this would happen were dramatically enhanced through the use of any number of inducements, usually in combination with each other. One way to draw the attention of small feeding flocks during spring and fall was to imitate the calls of the males. The hunter could produce the sound in his throat, but only at the risk of making swallowing painful. A more popular approach was to use two blocks of wood with a silk band sandwiched between. This "call" was held in the teeth and blown as one would a blade of grass held taut by the thumbs. An experienced caller would know both how to change the tone by varying the pressure on the blocks and when such a change would be most appropriate.[32]

People had long recognized that passenger pigeons were partial to salt. Although scarce in inland locations, salt springs made superb netting grounds. Several were known in Michigan, one of which by the White River was rented out for $300 a season. An even-better-known spring was in Benzie County. Discovered by pigeoners in 1870, it was called simply Salt Spring. Where the mineral-rich water gurgled to the surface, a mound had formed that sloped downward to cover an area of thirty or forty feet. During nesting years, pigeons by the millions congregated on this small plateau. Again the owners allowed netters to use the spot for a fee, but it was worth it as hundreds of birds at a time would be caught.[33]

Two pigeoners in Pennsylvania, F.E.S. and his companion, attempted to create their own salt spring to lure birds during the 1880 nesting near Sheffield. They selected a remote site with a deer lick and began clearing the vegetation in a rectangle of sixty by a hundred feet. "The muck was six feet deep, so we put in a good floor of poles and brush to prevent a trip to China." They spaded the top eight inches of the soil; dried it; shoveled the prepared soil on top of more poles so the bed would have a solid floor; and folded in five barrels of salt, ten pounds of sulfur, and a pint of anise oil. F.E.S. was particularly anxious to avoid the wet beds, or "old mud bed where the net, birds, and all went out of sight, and the birds were ruined for shipment." The innovation succeeded admirably and was soon adopted "by all first-class netters."[34]

In Maine, where rye fields were common, pigeon netters often prepped a different kind of baiting area, known as a dry bed, versions of which were used widely throughout pigeon range. The netters removed the stubble from a patch ten or twelve feet wide and fifteen or eighteen feet long. The dirt was leveled and took on the look of a vegetable bed. If no small trees were present, they would be brought from elsewhere and stuck in the ground to serve as perches. Finally, ample seeds would be laid out in rows.[35]

Grain of various kinds was often used to bait the pigeons. Corn drew their attention, while finer grain such as buckwheat would keep them feeding to encourage even more birds to alight. A farmer in Wisconsin did not specify what allurement he used, but he said the higher the quality, the more pigeons he would catch. Another pigeoner working in Pennsylvania found that adults had seemingly tired of grain after the squabs hatched, so he was forced to seek a more appealing substitute; angleworms proved to be just what the pigeons ordered, and by switching offerings he caught thousands of birds.[36]

Eliza Tucker, who lived near a large pigeon rookery in Richland County, Ohio, left behind several specific recipes for "Pidgion Bate" in papers dated January 10, 1826. One of them calls for the seeds of fennel, anise, and fenugreek pounded fine and boiled with "alwine." Add two grated potatoes and let the mixture stand covered for twelve hours before placing it as bait. A second recipe calls for boiling sassafras with wheat and provides some critical final details to ensure that the effort to attract birds is successful: "When you make your bed—to bait Pidgions—make it level—and smoothe—don't spit about it—nor make water—nor handle guns—nor Powder."[37]

As important as any other part of netting was a supply of live passenger pigeons for bait, Judas pigeons if you will. Three categories of pigeons were used: fliers, stool pigeons, and dead ones. Ensconced in the bough house, the netters would keep their eyes to the sky waiting for a flock to approach. At the right moment, the fliers, legs securely fastened by a cord sixty or more feet long, would be tossed into the air to entice flocks of foraging pigeons. Ideally the tethered pigeons would descend straight down without fluttering. Simultaneously, the pigeoner would be working the stool pigeon. The stool device was a stick about three feet tall that was pierced by another rod usually closer to six feet long near its top that pivoted like a teeter-totter. At the exposed end of the cross rod was a padded circular platform. Most often the platform was wood, although one fancy model used the bail of a pail covered with woven string. The bird would be fitted with leather or yarn bootees that were attached to the platform. This footgear could be tightened sufficiently to keep the stoolie from escaping but were of pliant enough material to prevent injury. Connected to the longer rod was yet another cord leading to the netter, so he could raise and lower it. Rapid lowering forced the stool pigeon to hover, making it look as if it were landing.[38]

Stool pigeons occupied a unique position in human/pigeon relations. Trappers would keep some number of their catch alive to find a few that would make good stoolies. One netter complained that he and his father nearly exhausted their supply of cooped pigeons before they found suitable birds, while another said that only one out of fifty pigeons would qualify. Males were generally preferred because they were larger and more brightly colored and thus more apt to be seen by passing flocks. A lot of care and attention went into these breathing decoys, and a bird with a good record could command a price of $5 to $10 or even more. The candidates were fed

Passenger pigeon net and stool held by Bob Currin, curator of the Coudersport Historical Society (Pennsylvania). The net was made by Aaron Robinson and was part of the society's collection when it was incorporated in March 1919. The stool originated with Earl Crane. Photo taken by author with permission of Bob Currin; additional information from Coudersport Historical Society

by hand to make them comfortable with people, and they underwent a rigorous vetting. The blindfolded bird in bootees balanced on the pigeoner's finger began his exercises: "The hand raised slowly and dropped quickly. As the bird drops, [the] wings are outstretched, quickly recovering as the hand stops, and this is repeated a number of times. Every motion is carefully watched and the action of the bird soon determines whether or not it will do for netting purposes."[39]

One aspect of the stool pigeons that has received a lot of attention is that they were temporarily blinded during the netting season so they would be immune from distractions: the sight of an incoming flock might cause premature movement that would alert the wild birds. The pigeoner would pierce the edge of one lower eyelid with a fine needle and silk thread, then go up and over the head to connect the lower lid of the opposite eye. The two ends of the string would be pulled to ensure the eyes were covered, then twisted together on the crown. One experienced netter said he never had a bird flinch or bleed during the procedure. Over several years the holes would become permanent, like those in human ears punctured to accommodate rings or studs. A few birds remained calm enough as to not need the fixing.[40]

Even while eradicating the species, some pigeoners became quite attached to particular stoolies. There was, for instance, the old trapper Jim and his Maggie. W. W. Thompson worked a nesting with Jim in Potter County, Pennsylvania. As one day ended with the capture of about two dozen birds, Jim enthused how well Maggie and the fliers had done in making that last small triumph possible: "There . . . you have seen as fine working of fliers and stooler, and as pretty a call of distant birds as you will ever see." Many of the pigeoners would gather in the evening in the basement of the Coudersport Hotel to talk of killing and other things, and to break in new birds. Probably over spirits a few nights later, Jim told Thompson that he was looking to replace Maggie. Many a stoolie had its career cut short by a hawk or other agent, and so Thompson immediately asked if she was dead. "Not that I know of, I hope not," replied Jim. He explained that he'd let her go, despite Thompson's offer to buy her a few days earlier for $10. She wasn't acting right, had a reduced appetite, and lacked the energy that was her wont. Ideally he would have rested her in his coop, to give her a chance to get over what was ailing her. But it might be months before he would be home. "I never kill a bird that does good work for me," he continued, "but turn them loose, and surely Maggie has earned her freedom . . . I sincerely hope she will escape the hunters and netters, and live as long as nature allows a pigeon to live."[41]

It is easy to imagine the netters furiously tugging the various cords in front of them like frantic puppeteers. The goal was to replicate the feeding behavior of wild pigeons, whereby birds at the rear of the flock would constantly be leapfrogging to the front to create the familiar wavelike action. Often the airborne pigeons would telegraph their intentions immediately and "lower their heads [to] come down." At other times the flock might go a half mile before starting to turn to scope out the scene further. If the first try by the netters failed, there would be additional forced ascents and descents until the coveted flock either passed or decided to join the Judas pigeons. To make the beds seem even more hospitable, trappers would seed them with dead pigeons, propped to look like feeding birds. But the incoming pigeons were wary, so everything had to appear copacetic; the tethered birds could not flutter or act in any other alarming way, nor could the bed display blood or feathers.[42]

The number of pigeons that could be netted at one time was astounding. At a nesting near Beekmantown, New York, in 1851, there were catches of a hundred dozen and eighty-five dozen each. Netters would not even bother springing their traps if the likely haul was less than forty or fifty dozen.

Forty bushels of corn lured a huge flock of pigeons to a trap in Wisconsin, which yielded thirty-five hundred birds in one catch. Perhaps, at least in part, because some of the last big nestings that took place there were so well documented, Michigan was the scene of some eye-popping single-haul totals: 300 dozen at a salt spring; 132 dozen at a bait bed (that's what was kept; additional birds escaped); and the precise total of 109 dozen plus eight claimed by Dr. Voorheis in Benzie County. Over fifty thousand pigeons wound up in the nets of one three-man team who worked the 1878 Petoskey nesting.[43]

A huge catch could be both a blessing and a bane. The strength of enough birds could lift the net, allowing many to get away. Usually the problem could be handled through fast work. Some netters kept rocks or additional poles by their sides to weigh down the net edges, but most often they leapt out of their makeshift shacks and started killing pigeons. The easiest targets were the birds that poked their heads through the mesh. In Pennsylvania it was popular to slay them by pinching their heads or necks with the thumb and fingers. After a while that would become tiresome, so flat rocks became the killing implements of choice. But clearly rocks weren't the answer either, so good ol' American ingenuity came to the fore when James V. Bennett invented, patented, and used a special kind of long-nosed pincer that did not close all the way, but enough to break pigeon necks. This device was not only more restful to the hand than other killing methods, but was said to have "effectually reduced the cruelty at the wholesale butcheries to a minimum."[44]

Another killing approach relied on a different part of the netter's body: the jaws. While some aimed to bite the pigeon's skull and crush it, others were more surgical and with practice could dislocate the neck bone without damaging the skull. One pigeoner complained that he chomped down on so many pigeon heads, his teeth became loose. (With the frequency of gum disease in the nineteenth century, that is quite plausible.)[45]

But at least one hunter balked at the oral attack: it was obviously a matter of taste. Novice Edwin Haskell shared a bough house with an experienced pigeoner. Their fliers and stoolie performed admirably that day, and a large flock of pigeons swooped in to investigate. At the perfect moment, the trap was sprung and the net scooped up a large percentage of the flock. But with so many struggling birds, Haskell had to launch himself onto the net to prevent the captives from escaping. He discovered, however, that even his weight and strength were insufficient to thwart the desperate pigeons: "We could not let go of the net to kill the birds with our hands—what, then, was to be done? The old pigeon catcher who had sprung the net decided quickly, by setting an example and yelling to me: 'Bite their heads! Bite their heads! Do you hear?' 'Not for all the pigeons in the world,' I replied. 'Pshaw! Don't be squeamish! See how it is done!' he called out impatiently and went on crushing the skulls . . . I could kill pigeons with a gun without any compunction. But crushing the skulls of live birds between my teeth! Faugh! It makes me shudder to think of it."[46]

CHAPTER 6

Profiles in Killing

> There is a peculiar charm about pigeon shooting; it captivates nearly everyone who participates in it. To-day we kill ten straight, and rejoice because of it; to-morrow we kill seven out of ten. Now we are dissatisfied and try again.
>
> —WILLIAM LEFFINGWELL, 1895

The great fun that naturally flowed from killing passenger pigeons was evidently not enough for some. They wanted competition and sought ways to turn the slaughter into a game. One way to do this was to assign points to the various animals that might be encountered in an area and then divide into teams. Each team would go forth early one morning and return in the evening with their load of game. The team with the highest number of points would win. One such contest was held in Berrien County, Michigan, in October 1878. Black bears were the most valuable at 300 points, followed by deer at 200, and otter at 100. Ranking dead last was "blackbird," worth but 1 point. Mud hens and small rails rated 3 points each. And passenger pigeons brought 10. The newspaper article did not report who won, but that team feasted on a "fine oyster dinner" paid for by the loser.[1]

Far more significant to the passenger pigeon story was trapshooting. In places across pigeon range, gentlemen would gather to shoot birds and place bets on the results. (Gambling contributed greatly to the popularity of these contests.) The large urban venues tended to draw the more refined sorts, who could afford to play, as the cost of quality pigeons and the upkeep of facilities was high. Amateurs were warned to stay away from shooting matches "unless [they] can afford to lose." Even if you wound up amassing more points than your opponents, the prize might still be less than what you would have paid for the privilege of shooting. People usually only entered the competition for the fun. This type of killing eliminated the rigors of traveling to remote hamlets and braving unseemly weather far from one's favorite edibles and beverages. This also removed the pigeons from any ecological context, transforming them from productions of nature to mere toys, although many, if not most, of the shooters had started out as field hunters, so it did not matter to them where they bagged their birds.[2]

Even supporters of pigeon shoots conceded that "field shooting is on a much higher plane than pigeon shooting could ever reach." Pigeon shooting was merely an exercise of one set of skills, claiming nothing more, although the activity might improve the amateur's use of the gun. If a field shooter relied on a dog to find his prey, he would lose the higher ground and become no different from any trapshooter. Although a good field shooter could always become a good trapshooter, the reverse was less likely to be the case. But in their own defense, the trapshooters claimed that their activity brought forth better than most anything else a man's inner qualities of "character, coolness, and determination." Merit, though, had nothing to do with the continued existence of these pigeon grinders. What kept them going until no birds were left was that those who pursued this avocation, at least in the cities, tended to be rich and powerful.[3]

Trapshooting began in England, but before there could be such a game, there had to be guns that discharged soon after aiming. These did not exist before 1790 or so. The first shooting club in the United States was established in Cincinnati sometime between 1825 and 1835. But as guns improved, so did the popularity of trapshooting. By 1848 "the sport . . . was pursued in a first-class style of excellence" on both sides of the Atlantic.[4]

In Europe the target of choice was the "blue rock," a cultivar of the rock pigeon that was smaller than other strains and could fly fast even as it zigged and zagged. In the United States, opinion was mixed as to which pigeon made the better quarry. Passenger pigeons flew faster but usually in a straighter trajectory. Additionally, the domestic pigeons, being the product of husbandry, did not have to withstand the ordeal of lengthy travel and often squalid confinement experienced by the wild birds. Both pigeons were utilized, but the cheapness of the passenger pigeons during their period of abundance often trumped other considerations. There were, however, admonitions that only adult passenger pigeons in good physical condition could provide satisfactory sport. Still, on occasion birds were so caked in excrement from being tightly packed over long distances they could barely fly, if at all.[5]

As important as the birds themselves were to the competitions, the devices or traps used to thrust them skyward often made all the difference in how they performed as targets. There were two basic types of traps. The first was the ground trap that consisted of a box with sides and a top that fell away to release what were mostly rock pigeons. A variant was a half-cylindrical container placed in a small hole on a horizontal pivot. When a rope was pulled, the pigeon was "scooped" into the air. But this enabled the bird to fly low without restraint in any direction, allowing many to escape the shooters, much to their dissatisfaction. Hence, these were not used much.[6]

For a while the ground trap was largely supplanted by the second type, the plunge trap. Here the bird was placed in what looked like the standard ground trap or one shaped like a pyramid (the pyramid design was made by Parker Brothers), but the trap was on top of a spring-loaded plunger. When the plunger was released, it shot up, the sides and top would give way, and the bird was propelled into the air. This rendered the pigeon little more than an inanimate object during that brief period when its movement was dictated by the force of the plunger, so the hunters fired quickly. Indeed, John Brewer, one of the great shooters of all time, revealed the key: "The secret of pigeon shooting is to kill the birds quickly, they must not be permitted to become hard birds; the quicker the first barrel is fired the better, and second must follow before the bird is forty yards from the shooter." Only birds that fell within a certain distance registered points for the shooter.[7]

But these plunge traps could be manipulated by the trap operator in ways that favored one shooter or another. Often the plunge traps were used in what were called "find, trap, and handle matches," where each shooter had an accomplice who handled the birds and operated the plungers. This allowed not only shenanigans in the way the trap was run, but also in the treatment of birds. Cutting off the tail tips or clipping the toes would ensure that the bird took off at the earliest moment. In these types of matches, the handler was often deemed as important to the outcome as the shooter, for the "betting was governed very often solely on the handler."[8]

Sometimes the passenger pigeons and other types did not want to fly when they were released. To scare any recalcitrant birds into action, a fellow named Rosenthal invented a mechanical cat that would suddenly leap to its feet with the tug of a cord. More often than not, though, the use of such fancy contrivances to foster the desired flight was rejected in favor of cheaper methods. Probably most common among these was the use of club-wielding men, but other approaches were tried as well: tearing away back feathers and applying cayenne oil to the exposed skin; puncturing the bodies and toes with pins; and blinding one eye with sticky plaster, tobacco juice, or turpentine so the bird would fly in circles.[9]

During the Civil War pigeon-shooting meets went the way of peace, but returned with the cessation of hostilities. By this time, there was a national market in passenger pigeons, and the birds were being exploited as never before, leading to a ready and cheap supply. Shooting clubs and well-publicized matches spread across the country from Texas to the East Coast. These were popular in Ontario as well, but they were smaller in scope given a generally less affluent group of participants. By the end of the 1870s, pigeon shoots had become a mania across pigeon range. As one writer put it, there were shooting matches in every little town and a champion at every crossroads. Looking at shooting matches across the 1870s and 1880s, Edward Thomas stated that during the April-through-September passenger pigeon season, "probably half a million were used and twice, perhaps three times, as many of the common or domestic variety, particularly in the country towns and when the supply of wild was exhausted." Based on the totals published for one year in just one hunting journal, *Forest and Stream*, it was estimated that, in 1880, 62,868 passenger pigeons were shot at and 44,668 killed.[10]

The largest contest ever held in the south occurred in Dallas in May 1880. Five thousand pigeons "are to be slaughtered during the tournament." But the sponsors, the state sportsmen's association, had to hustle to obtain the promised number. Twenty-five hundred arrived one day from Chicago, while seventeen hundred more came a week later. Presumably successful delivery of the remainder was made before the commencement of the shooting. The following year's tourney, held at Denison, also featured five thousand passenger pigeons, but there seems to have been a misunderstanding with the supplier, and some of the expected birds never arrived. These were birds from the big nesting near Atoka, Oklahoma. The Sparta, Wisconsin, nesting generated the five thousand birds that were shot at the 1882 meeting of the Austin Gun Club.[11]

A Texas match held at Hockley, near Houston, in 1884 offered two hundred freshly caught pigeons. But of greater interest, it featured an innovation: "Mr. Ellis [the organizer] has his dogs trained to retrieve from the traps, a novel feature in trap shooting, and one that, universally adopted, would spare the pain incident to many an average small boy being peppered with small shot." Ellis, exercising his Texas flair for embroidery, commented that soon his dogs would also be able to place the live birds on the traps.[12]

The St. Paul, Minnesota, Sportsmen's Club was known for being one of the wealthiest clubs, so they did not scrimp at their shooting matches. They were close to the big nestings in Michigan and Wisconsin and often visited the sites to make arrangements directly with netters. In 1874 they announced plans to build a coop that would accommodate several thousand pigeons. Their 1878 shooting contest offered contestants ten thousand birds to shoot.[13]

Chicago became a center for pigeon shoots because several large dealers had their headquarters there. Dexter Park, on the South Side, was a popular venue. The Kennicott Club met there in August 1872 for a match that drew forty-four contestants vying for prizes. Each shot at twenty birds, and after the first round three shooters had killed nineteen. A second round was called to settle the tie. Each man was to shoot ten pigeons. This eliminated but one, as two reached the identical scores of eight. These two then faced off, and each was given ten more pigeons to show off his prowess. But again they each bagged the identical number—seven. Yet another round of ten was ordered, and this time a clear winner emerged, as the score was eight to four. It was said to be the most exciting match at Dexter Park in years. Five years later, the same venue would host "a grand tournament" using five thousand wild pigeons. A few hours to the southwest, the Illinois State Sportsman Association held its annual meet in Peoria in 1879 and had fourteen thousand pigeons on hand for the festivities.[14]

Like every other commercially motivated sport that seeks to expand its popularity, trapshooting began cultivating its stars. A special medal for "Champion of America" was offered to the best shooter. If that person could hold the trophy for two years, it was his to keep. The first competition for the medal was held at Mark Rock, Rhode Island, on April 7, 1870. Six shooters were presented with thirty-five birds each, and a point was earned for every bird shot. Thirty-two was the winning score. The championship changed hands several times over the first year and a half of the contests, but then there emerged "a champion of champions," who was to reign as the best shooter of all until the ravages of age eventually robbed him of his superior gifts. His name was Adam Bogardus.[15]

Bogardus was born and raised in Albany County, New York. Even as a teenager, he excelled at shooting snipe and bobwhite, limiting himself to a daily take of twenty-five brace. Most of his life, though, was spent in Elkhart, Illinois. To the south was Christian County, and there in one three-month stint he made it a personal crusade to reduce the avian life of that prairie district: he shot over three thousand birds, including prairie chickens, geese, ducks, whooping cranes, golden plover, and Eskimo curlews.[16]

Adam Bogardus

For eighteen years he stalked the fields and marshes of the Midwest killing birds before he participated in his first pigeon shoot in 1868 in St. Louis. (He had never even seen a pigeon trap before.) He was so successful that a sponsor arranged a match between him and a vaunted shooter from Detroit named Gough Stanton. Stanton agreed to come to Elkhart, but he brought a plunge trap of a kind that was new to Bogardus. Despite that, of the fifty birds each had to shoot, Bogardus won forty-six to forty.[17]

He then engaged in a series of duels with the king of Chicago shooters, Abe Kleinman. Kleinman and his brothers ruled the Calumet area, the large complex of marshes on the city's southeast side that was nationally famous for its waterfowl. The first battle royal between the two also involved fifty birds each, shooting for $200 a side. It was a trap and handle match, and Kleinman had the better handler, for Bogardus complained that he was saddled with an accomplice who did not appear to "know an old bird from a young one." The betting grew more intense, with Bogardus placing additional money on the proposition that he could kill forty-six of his birds. This he accomplished with no birds to spare, but Kleinman missed only one. When they later vied for the championship of Illinois, Bogardus prevailed by a single bird.[18]

Their next match was something of a novelty. Bogardus agreed to shoot from a moving wagon four yards closer to the trap than Kleinman, who was stationary. Bogardus attributed his victory to the years he had been shooting "from a buggy at plover, grouse, and geese." This practice had made him "very quick and effective." Kleinman and Bogardus competed against each other in another unusual match a few years later, with Kleinman and three other crack shots each having fifty birds to shoot. Their combined scores would be pitted against what Bogardus could do on his own shooting at two hundred birds. Again Bogardus had the better of it, winning 178 to 176.[19]

In 1872 Kleinman and Bogardus competed against each other in yet another memorable match, held at Dexter Park in late September. It was $500 a side and each shot at a hundred birds. Bogardus again won, scoring 85 to Kleinman's 81. The weather was awful, with a furious wind making it hard to shoot and unpleasant to watch. But over five hundred did show up, including Lieutenant General Philip Sheridan and Lieutenant Fred Grant (President Grant's son). As proof that the audience and participants of these matches were hearty, another match at Dexter Park occurred on New Year's Day.[20]

Bogardus traveled to St. Louis in May of 1880 to compete against that city's top shooter. He brought along his entourage, which included his usual rival Abe Kleinman. Three thousand people watched Bogardus prevail once again, defeating the local favorite 86 to 83. The passenger pigeons originated from Michigan "and were in all sorts of fix, some very lively and some very weak."[21]

Bogardus was open to new types of contests and showed his flexibility in another Chicago match. In this instance he was shooting for $1,000, plus numerous side bets. But he was not going against another person. His goal was to kill five hundred passenger pigeons in 645 minutes with one gun that he had to load himself. Here, too, he was victorious, having gone through all the pigeons with almost two hours to spare. He missed few birds and in one streak killed seventy-five consecutive pigeons. No doubt the five hundred birds he killed in practice kept him sharp for the actual contest.[22]

As befitting New York's standing in the late-nineteenth century as the country's center of commerce and culture, the pigeon shoots staged there were the biggest anywhere. The New York State Sportsmen's Association put on splashy and elaborate tournaments. Their 1874 extravaganza claimed forty thousand to forty-five thousand passenger pigeons. Another of their matches held in Syracuse in 1877 used twenty thousand. And then there was the contest put on as part of the Sportsmen's Association annual meeting in June 1881. It took place on Coney Island and involved twenty thousand to twenty-five thousand birds.[23]

The shoots in New York also provoked the first organized and effective attempts to stop the slaughter of passenger pigeons. It was not based on fear that the birds would become extinct, but rather on the cruelty of the game, which fostered revulsion. Some have argued that once consumption shifted to the urban areas far from where the pigeons lived, the birds ceased being living entities with specific requirements: "At a trap shoot, people who used pigeons intensively lost track of pigeon natural history. They also lost track of their own connections to pigeons—and of the consequences."[24] But no evidence suggests that rural people were any more disposed to protect pigeons than anyone else: they saw the birds as a resource that provided them with extra income during pigeon years.

It was not someone who had experienced pigeon years, but rather a rarefied New Yorker named Henry Bergh, and his minions, who first launched a concerted attack on the wasteful slaughter of the trapshoots. Bergh was born in New York City in 1811. His father was a wealthy shipbuilder so Henry was allowed to wander through life with greater freedom than is usually granted. He attended what was then called Columbia College (now of course University), but stayed only a year before embarking on a five-year sojourn through Europe. He fancied himself a writer, authoring plays and poetry that were apparently never highly regarded by others. When his father retired, Henry and his brother shared the business, but Henry's heart wasn't in it. Upon his father's death, Henry sold his share in the company and spent the rest of his days unburdened by the need to make a living. Sometimes this leads to a life of dissipation, but in other instances the luxury of being able to focus on issues beyond one's self leads to important accomplishments. Clearly, Henry Bergh falls in the latter category. In the words of a biographer, "few public figures have labored with a greater zest for battle, or a more flamboyant sense of the dramatic, than single-hearted Henry Bergh."[25]

In 1863, Bergh, along with his wife, Catherine Matilda, left for Russia, where he was to serve as secretary of the American legation in St. Petersburg, a position to which he had been appointed by President Lincoln. But for reasons that are not clear, they did not stay long. By then, however, he had cultivated a deep and abiding aversion to cruelty. For the rest of his life he fought against brutalizing animals through the organization he founded in 1866, the American Society for the Prevention of Cruelty to Animals (ASPCA), the first such society in the western hemisphere. He would say, "Mercy to animals means mercy to mankind." He meant it, for nine years later he cofounded the Society for the Prevention of Cruelty to Children.[26]

The last four decades of the 1800s offered an exciting variety of competitions to the devotees of rat baiting, dogfighting, bearbaiting, and even exotic contests between such animals as lions and bulls. On one cold night in

February 1867, for example, New York cognoscenti enjoyed a bout between a local favorite, the bulldog Belcher, and a visitor from Boston, the bitch Venus. Backers of each put up $500. After fighting for thirty minutes, Belcher made the home crowd proud, leaving Venus's "head a mass of blood, her jaws, jagged and torn, dripping a purple fluid, one ear torn away, and altogether not a kick left in her." One venue in Philadelphia, unable to obtain dogs from New York, offered a show pitting twenty-four large rats against "a bull-headed little man," who plowed into the milling rodents, grabbed them to his mouth, and broke their necks with his jaws. He yanked rat hairs from his teeth and cleansed his palate with a slug of whiskey. While dogfighting still goes on, it has been illegal for a long time. Bergh's challenge in ending these contests in New York was enforcement. But his campaign against the pigeon shoots in New York required a preliminary step: he first had to get them outlawed.[27]

Bergh started his crusade against the pigeon-shooting matches in 1869 by prosecuting their organizers under the New York cruelty law and a municipal ordinance. In response, the specific match he was trying to shut down moved across the river into New Jersey. This gave pro-shooting forces time to organize and they issued a warning to Bergh: if he took them to court, they would get an injunction and gut the law that enabled him to prosecute cases involving other forms of cruelty. His confidants persuaded him not to press the shooting issue so he backed off reluctantly: "I still indulge the belief that the day will come when the 'sport' in question will be substituted by a more humane pastime."[28]

Some powerful people who inhabited the same world of privilege as Bergh supported the contests. One was the newspaperman Robert Roosevelt (Theodore Roosevelt's uncle), who said of Bergh, "He is about the best intentioned and least practical man in the community. The idea of cruelty in field sports has long been exploded." The shooters made a big point of claiming how careful they were in preventing the living targets from being pained. The bigger matches had armed men standing on the edge of the field to dispatch any wounded birds that might have eluded death by the competitors. Boys were around to catch the birds no longer able to fly, while additional gunners stood outside the gates ready to shoot any bird that might make it that far. Promoters claimed that all of these precautions prevented unnecessary suffering and virtually assured that few birds released for the meets survived the gauntlet.[29]

Bergh did have his supporters, though, with one newspaper pointing out that his critics supported the reforms on the treatment of draft animals, but when it came to stopping cruelty in a sphere they found entertaining, their views suddenly changed. Through constant effort, public attitudes toward the shoots began to shift by 1881. Already, such states as New Hampshire, Rhode Island, and Massachusetts had outlawed the meets. Pro-Bergh New York papers headlined their views on a proposed match: "A Brutal Exhibition"; "Slaughtering: A 'Gentlemanly Sport' "; and "Protect the Pigeons."[30]

The *New York Times* ran a satirical refutation of the argument that pigeon shooting prepared marksmen for military service and therefore enhanced national defense:

> It is generally supposed that the pigeon is a harmless and inoffensive bird, but who can tell what depths of ferocity lurk beneath its smooth and plausible plumage? We should be in a nice position were an army of millions of pigeons to suddenly fall upon us and begin to peck our eyes out. While we heedlessly ignore the possibility of such a calamity, brave and prudent pigeon-shooters are acquiring the skill which will enable them to defeat an invading host of infuriated pigeons . . . But if it is amusing to kill a bird, how much more interesting to kill a man! Men shooting would prepare sportsmen to defend their country against an invading army of Europeans.[31]

The New York State Sportsmen's Association's annual meeting opened at Coney Island on June 20, 1881, to all the pomp and splendor that usually marked such events. The first day was devoted to business, but the actual meet started on June 21. Ten plunge traps stood twenty-one yards from the shooting platforms. The contestants came forward in "teams of twenty each and shot alternately in squads of 10 each." The various member organizations ran hospitality tents that dispensed copious amounts of liquid refreshment, food, and cigars. Housed in a temporary shed nearby were stacks of crates filled with the thousands of pigeons that would provide the entertainment. Despite the cloudless sky and cooling breeze, few spectators were in attendance. At twelve thirty a closely watched party of three bought tickets and made it as far as the grandstands before they were stopped. Bergh had been recognized and was confronted by Abel Cook, president of the state association. Bergh demanded that he be allowed to inspect the premises to ensure that the birds were being treated in a humane manner. Cook said it was obvious nothing untoward was going on and denied Bergh further access. Bergh retreated to the grandstand and lingered an hour scanning the proceedings with binoculars. He then left, but not before conveying to Cook his intention to go straightaway to Albany to make sure such a meet never occurred on New York soil again. The law that emerged later that year became a model for the country.[32]

As for this last meet in the state, the *New York Times* noted that a boy broke his arm by stumbling into a ditch as he chased a wounded pigeon. On the last evening, "carelessness of the person in charge" caused a fire that destroyed one tent and a number of valuable guns housed inside.[33]

Shooting matches continued in the places where it was still legal. As the passenger pigeons became more and more difficult to obtain, those marksmen who felt deprived if they could not kill something began using substitutes. Besides rock pigeons, a host of other critters were enlisted as targets, including house sparrows (the weaver finch formerly known as the English sparrow), purple martins, blackbirds, bats, mallards, rabbits, and bobwhite. The quail would apparently fly in a straight line, and a likely score out of ten was only six or seven. Snow buntings, "as game little fellows as ever lived," proved the most challenging, though, especially when snow was on the ground. A white target against a white background meant that a shooter would be lucky if he could see, let alone nail, half his birds.[34]

Eventually, inanimate objects replaced live ones in most places. First there was the glass ball, one variant of which was developed by Captain Bogardus. But that was replaced by the "clay pigeon," patented by George Ligowsky in 1880. Ligowsky went through forty different configurations before settling on the saucer-shaped target that is used today. Bogardus gave his endorsement: "Clay pigeons are by far a Superior Article for the sportsmen . . . This new invention will largely fill the void in trap shooting made by the scarcity of the wild pigeons."[35] Indeed, clay is still going strong.

ETTA WILSON: A CHILDHOOD OF PASSENGER PIGEONS

Shooting passenger pigeons was an important part of Etta Wilson's heritage. Her later revulsion to what she had witnessed and abetted as a child led her to a distinguished career in ornithology and conservation. She worked tirelessly on behalf of the National Audubon Society and Bureau of Biological Survey, researching such topics as migration and the spread of house sparrows, as well as lecturing to over twenty thousand people a week. A newspaper

Etta Wilson. Courtesy of Connie Ingham and Cindy Laug, Grand Valley State University

reporter for a long time, she left behind a quantity of writings documenting the varied chapters of her life. And a most extraordinary life it was.[36]

Her paternal grandmother was Kin-ne-quay, the daughter of the great Odawa (Ottawa) chief Joseph Wakazoo. She herself attained distinction as a healer, providing relief to many patients, although unable to save her husband and six children from smallpox. Tribal elders pressed her to marry again to help maintain her royal lineage. She resisted strenuously, saying no one available was worthy of marrying a Wakazoo. In her forties, after great trepidation, she finally relented to wed Nayan Mi-in-gun, two decades her junior but from a family as distinguished as her own. Their union produced one son, Payson. Nayan Mi-in-gun died in a hunting accident, so it would be just the mother and child.[37]

Congregational minister George Nelson Smith and his wife, Arvilla Powers Smith, left Vermont in 1833 and established a mission in Kalamazoo County, Michigan. George Smith was eloquent and persuasive, and visiting Indians began talking him up to others. In 1837, a contingent of Odawa co-lead by Wakazoo traveled from Emmet County (just south of the Mackinac Straits) to meet the reverend. They were so impressed with each other that Smith committed his life to "the spiritual and temporal welfare of the Indians." Wakazoo and his followers, in turn, relocated to Allegan, Michigan, where they formed a village devoted to Smith's teachings.[38]

The Smiths had a number of children, including their eldest daughter,

Mary. She and Payson attended classes together at the village school and both spoke fluent Odawa and English. He, the son of a princess and the grandson of a war chief, and she, the daughter of the village's leading white citizen—how could it be otherwise? They fell in love. Kin-ne-quay was devastated when she learned and rushed to Rev. Smith in a "furious rage": "I will not let my son marry a white woman and learn to be a white woman's man, to lie, to steal, to cheat, to swear, to go out and get drunk and come home with strange diseases." Smith was none too keen on the developments either, and Arvilla remained opposed until she died. But the community elders supported the marriage, and Smith realized that he couldn't very well obstruct the strong desires of the two young people without looking like a hypocrite to his parishioners. Like parents everywhere before and since, Kin-ne-quay performed some mental gymnastics to overcome long-held beliefs and finally assented to the arrangement. On July 29, 1851, Smith presided over the nuptials of his daughter and Payson Wolfe, who now used the anglicized version of his name. In his diary, Smith noted that "the occasion was pleasant."[39]

Payson and Mary Wolfe had thirteen children, including Etta, who was born in 1857. Their farm occupied a narrow stretch of hilly land that overlooked Grand Traverse Bay. Most of the property supported a thick woods, dominated by beech, maple, cedar, and hemlock, but with some open patches, including a four-acre pasture.[40]

Etta and her family looked forward to the April arrival of the pigeons. Vast flocks streamed to a favorite breeding area not far away, where they would spend the spring and early summer caring for their young. On occasion the Wilsons witnessed a reverse migration, with strings of pigeons moving south. She surmised that these were birds aiming for a more northerly destination that inadvertently wound up on the peninsula. When they reached the tip, they preferred to backtrack rather than cross the bay.

When it was pigeon season, her father would shine, for as befitting the progeny of Odawa chiefs, he exceeded all others in his prowess as a pigeon hunter. Payson's goal was to shoot a thousand pigeons before breakfast. The children rushed to the house with a basket as soon as it was filled. The bodies were dumped and they headed back to Payson, where on a good day another filled container would await. This running back and forth continued until the quota was reached. At the house Mary and some of the older children would sort the birds and dress the ones for the morning meal. Payson had a knack for knowing when he'd reached his number, at which time he would usher the children homeward.

The neighbors recognized the Wolfe farm as the premier shooting spot in the vicinity. Virtually every male old enough to handle a gun in the nearby town of Northport would call on the Wolfes to try his hand in the carnage. Many relied on the sound of Payson's gun to signal that pigeon-plunking had begun and it was time to climb the hill. While the neighbors killed many tens of thousands of birds each season, none were as adept as Payson.

The killing was bad enough in the spring when the adults returned, but later when the fledglings began flying, the fusillade proved devastating. Their response to the hail of bullets was to close ranks, seeking protection in one another. This merely made for a larger target. But Etta observed them react in an even more self-destructive way: "I have seen a flock of hundreds of inexperienced birds fly up the hill to and meeting the barrage of guns, falter, hesitate, turn and pass the entire gamut of the line thus exposing themselves twice to the attack, and return whence they came. If twenty-five individuals survived the double onslaught they were lucky."

When the birds exited the forest, they dipped low over the pasture and stayed just above the ground as they cleared the hill. Here not just the male neighbors gathered, but many females as well, for guns were not necessary to partake of the festivities. Clubs and rocks proved effective as well. Etta's huge family included a canine of bull/shepherd stock whose strength was as a watchdog, not a hunter. But even he could not contain himself when the pigeons swooshed over his head: leaping ponderously, he actually grabbed a bird out of the air!

Etta's first job was to retrieve the birds her father had only wounded. She snuck through grass as tall as she was to collect them. Etta struggled over these damaged birds. While holding the fallen pigeon gently in her little hands, she would study it, noting its "wild frightened red eyes and with one or both wings or legs broken or shot off, while its little heart beat in rapid tempo against [my] palms." She could not kill one herself, so she handed the prisoner to her father or a brother—a quick snap of the head and the bird would be stilled.

For reasons not stated, one of these pigeons enjoyed a reprieve. Named Partie, she roamed freely throughout the house as she recuperated from her bullet-induced injuries. Her principal mode of transportation was Mary, on whose shoulder she would perch as her ride performed domestic duties. Occasionally, the hitchhiker would "reach up to caress mother's cheek with her slick little head." Partie slept on a fresh bed of newspapers between two books on a shelf. Etta says the bird recovered completely save for the loss of a few feathers, but her eventual fate went unreported.

After breakfast there remained hundreds of pigeons to process. Some number became food for the family—the fifteen mouths could devour nearly fifty pigeons in a day. Other birds were salted for consumption during the winter. And whenever a neighbor asked, Mary generously replied, "Take all you want."

Most of the pigeons, though, were prepared for market. The family busied itself sorting the birds: the better ones bound by the legs and sold by the dozen, while others were crammed into barrels. The Leelanau Peninsula was free of railroads, so all the birds went out by boat. During pigeon season numerous vessels stopped nearly every day, and Payson sold most of the birds he had, either those in barrels, destined for Chicago, or by the dozen for consumption by the boat crews. These latter fetched anywhere from five cents to a dime, depending on the overall quantity available at the various ports. When prices were too low or the market too saturated to merit cash purchases, Payson would on occasion trade the birds for "baker's bread" (as opposed to the less desirable homemade bread). Twelve pigeons would bring one loaf.

Neither Etta nor her father ever visited the nestings that took place in northern Michigan. But her brothers did and they returned with tales of "horror": "Gnomes in the form of men wearing old, tattered clothing, heads covered with burlap and feet encased in old shoes or rubber boots went about with sticks and clubs knocking off the birds' nests while others were chopping down trees and breaking off the over-laden limbs to gather the squabs." Etta called it "an inferno where the Pigeons had builded their Eden."

Etta noted the yearly decline of the pigeons. Every year fewer birds stayed to nest, until 1878, when there were none. That was the year of the last great nesting that took place near Petoskey, across the bay to the north. Others also noticed that the spring flocks seemed less dense than they used to be. Few thought the birds would vanish altogether, but merely that they would find another place to gather and that the residents would be denied a source of food and commerce. But the wiser among Reverend Smith's Odawa congregation did complain to him that if the hunting did not let up, there might actually "be no more pigeons." He conveyed their concerns to the professional pigeoners, suggesting that they give the birds a respite every other year. "Bosh" is the four-letter word Etta uses as their response, though likely it was harsher than that.

Soon after the 1878 nesting at Petoskey broke up, Etta and her family woke one June morning to find the surface of Grand Traverse Bay and its shoreline covered with dead pigeons. It had been a fair night, free of wind and fog. Etta wondered if it was possible whether the birds, wrought with despair and weary of endless torment, had "flung themselves into the waters of oblivion." The Odawa gazed upon the windrows of pigeon corpses and harbored no doubt: "They have committed suicide. Their persecution was more than any living thing could endure."

DRAINING THE LAND OF ITS PIGEONS: SCIOTO MARSH AND MEANS PRAIRIE

The unpublished memoir of George D. Smith, describing the obliteration of a pigeon roost in Auglaize County, Ohio, is one of the most valuable in the substantial annals of the bird's history. Unlike accounts that focus on the clearing of forests as a factor in the bird's extinction, Smith points to the loss of wetlands as the critical development that sealed the fate of this particular roost. Passenger pigeons, besides all the deliberate actions taken against them, were here the inadvertent victims of federal law.

Smith was born in 1865 and grew up in Waynesfield, not far from the Scioto Marsh, where the pigeons roosted, and Means Prairie, where they foraged. He received two undergraduate degrees from Ohio Northern University and Ohio Wesleyan. Teaching science was his calling, and he devoted his career to it at various institutions, most especially Eastern Kentucky State Teachers College (now Eastern Kentucky University), in Richmond, where he spent twenty-five years teaching biology. He was also an accomplished photographer, and his nature pictures were widely published.[41]

Smith's earliest recollections were of continuous lines of pigeons flying over his house as they left the roost in the morning and returned to the sheltering woods of the swamp in the evening. The birds occupied the skies from four to seven or eight on both early and late flights. The heights they flew exceeded the range of firearms, although it did not stop the young Smith from trying. His shots skyward never hit any pigeons, but they caused the flocks to separate briefly before coalescing as before.[42]

The prairie occupied the headwaters of the Scioto River, seventeen thousand acres of lowland "largely covered with weeds, prairie grass, willows, and other moisture loving plants." Smith makes no mention of oaks or other nut trees, and since this was in winter, no fruits would have been available. Maps of presettlement vegetation place this in a region of ash and elm on whose seeds the birds were known to feed heavily. Downstream sprawled Scioto Marsh, twelve to fifteen miles long and two to four miles wide. Both sites were largely inaccessible to humans, and that is one attribute that held the birds. The one exception was in the winter, when the wet areas froze. But even then the danger of breaking through the ice kept all but the most intrepid visitor from entering what seemed to be a wasteland good only for pigeons.

To encourage the draining of marshes and swamps so they could become "productive," Congress enacted the Swamp Land Act of 1849. This statute applied only to Louisiana, but a year later Congress extended the provisions to additional states, including Ohio. Federal lands would be conveyed to the states, which would in turn transfer title to private parties for little or no compensation provided they drained the property. In this case, Limon Means received the seventeen thousand acres of prairie and quickly began to fulfill his end of the deal: "Mr. Means employed a lot of Irish laborers to ditch it, and before many months had passed there were many large ditches running all thru it from which gopher ditches ran out into the fields. The weeds and grass were burned off and during the following winter and spring, long strings of oxen could be seen turning the sod for a corn crop."

During this same time, the marsh, too, came under assault. A dredge was shipped downstream to Kenton, from where it was hauled by oxen to the marsh: "For nearly two years that dredge puffed and snorted in the bed of the river. People came for miles and miles to see it for it was something new in that community. In due time the river was opened up and the marshy land drained."

Suddenly, nothing protected the roosting and feeding grounds of the pigeons. Hunters invaded the marsh by droves, for the roosting birds could be killed with little effort. A neighbor named Jim told Smith about the happenings in the marsh: "Do you know that they are catching [pigeons] over there by the millions and hauling them away by the wagon and sled load?" He suggested that he and the Smiths pool their horses so they could rig up a wagon that could hold sixty bushels. Jim and his two sons along with three Smiths ventured out on a cold December day in 1877 to try their luck. Provisioned with lanterns, bags, nets, food, and clubs, they arrived at their destination at dusk. As they waited impatiently, they could see great crowds of people pouring in; this parade would continue into the night and included visitors from forty miles away.

When darkness eventually arrived, they went forth with blazing lanterns into the roost: "We had not gone far when we came to a small cluster of willows which was filled with wild pigeons. Our lanterns blinded them, so they did not fly until we had formed our circle ready for action. The word was given and we made the attack. Instantly, there was a sound of many clubs beating from all sides against those helpless birds. They fell by the score but a few of them fluttered away in the darkness and were lost to our view." Their few minutes of labor produced 114 dead birds.

The next willow stand was somewhat larger than the first, and they decided to employ their homemade net, almost eight feet high and fifty or sixty feet wide. Previously, they had used it to catch fish at nearby Lewistown Reservoir. Four members of the party encircled the birds with the net. They began clubbing pigeons within reach while frantic birds seeking escape flew into the net and fell to the ground. Eventually those birds, too, would be bludgeoned to death. In the end not one bird escaped, and they had added 225 pigeons to their total. By eleven that night they had filled sixty bushels with pigeons and were ready to call it a night. Their take hardly made a dent on the total population of the roost, but they were hardly alone: "The crowds could now be counted by the hundreds, and the lanterns could be seen throughout the valley darting back and forth like a great swarm of fireflies."

Smith and his friends spent the next day handing out pigeons to their neighbors. Such generosity was warmly appreciated, and the tale they told spurred over fifty of the recipients into launching their own raiding parties. The word continued to spread throughout the winter and into March: "Finally the number of hunters became so vast, they penetrated every nook and corner of the valley, and so the remnant of the pigeons left the Marsh and found a roosting place on the Prairie." But here, too, the birds were sought out and destroyed, forcing them to abandon this refuge as well. The birds stayed for a short while in nearby woods but soon thereafter made their exit: "Then suddenly as if the earth had opened up and swallowed them, they disappeared and never again with only one single exception the next winter, did we ever see a pigeon in that region."

PIGEON SOCIAL

Come for the pigeons, stay for the booze, bands, and balderdash!

Evadyne Swanson's admirable dissertation on the treatment of Minnesota's wildlife from 1850 to 1900 refers to the "social and convivial" element of passenger pigeon hunting. The accounts from Minnesota, she says, are rich in jocularity, while those of Wisconsin and Michigan consist mainly of "sordid tales of systematic slaughter." This difference is most likely due to the editor of the *Chatfield Democrat*, the paper from which all of Swanson's Minnesota examples originate. He obviously had an eye for the sarcastic and humorous, and his reports reflected that.[43]

Many of the *Chatfield* yarns center upon the liberal indulgence of distilled spirits. In the June 6, 1863, issue, the editor relates his encounter with a group of hunters from the nearby town of Cremona. They urged him to join them for the night, and it proved to be one of his more enjoyable experiences: "We ate pigeons cooked in every style, until 'we couldn't rest' without divers and sundry 'night-caps.' Take it all together it was a time long to be remembered . . . But we must say that we had pigeons enough; too much of a good thing is too much, and we believe the unanimous verdict of the party was that fat squabs, whiskey, and boiled eggs are a bad mixture, which refuse most decidedly to mingle together."

Two years later, the pigeons nested in the neighborhood again. In late May the "Pigeon Roost Camp Meeting" was planned for all those who enjoy "sport and squabs": "Preying will commence at early dawn and continue till early candlelight each day . . . Arms requisite for successful 'squabbing,' a long pole and high boots—provisions for the camping, whiskey, smelling salts, bread, salt, and a little whiskey." Given it had been a late spring, the squabs proved to be still a bit small so the camp meeting was postponed until June 3:

> Bishop E. D. Williams, Presiding Elders John A. Mathews, A. W. Webster and several other brethren loud on the amen of Winona, having arrived yesterday afternoon. Deacons Hyde, Broughton, and others equally pious, are expected this evening from St. Charles. A delegation of devoted Christians from Preston will also be present. As all of the brethren who will be in attendance are spiritually inclined, we presume much squabs will be devoured and that old devil, whisky, severely punished. Amen!

When the festivities had finally concluded, it was estimated that all the joyous preying claimed ten thousand squabs.[44]

Good humor also sprang from the breach in the britches of old squabber Isaac Day. It seems that the year before, in 1864, another squabbing festival drew folks from all over the region. The editor opined that "stewed squabs with whisky sauce [were] not bad to take if you knew when you have eaten enough." During the festivities, Mr. Day scampered up a tree to grab squabs, and his pants split to the great amusement of the gathered observers. This year everyone was warned to wear tear-resistant slacks. Day and his pants were given another chance. As he scaled the tree, cheering spectators waved flags, and to honor the integrity of his trousers, the band began playing "Yankee Doodle." All the while, the ladies "clapped their pretty little hands . . . crying out in the fullness of their overjoyed hearts, did you ever see such a day?"[45]

CHAPTER 7

The Tempest Was Spent: The Last Great Nestings

> The pigeon was no mere bird, he was a biological storm . . . Yearly the feathered tempest roared up, down, and across the continent, sucking up the laden fruits of forest and prairie, burning them in a traveling burst of life.
>
> —ALDO LEOPOLD, 1947

Whereas passenger pigeons once enjoyed an extensive nesting range, all of these last large nestings were in six states and one province: Wisconsin, Michigan, Minnesota, Pennsylvania, New York, Oklahoma, and Ontario. Unfortunately, little is known about many of them. This is particularly true with regard to Ontario, where most of the nestings have come to light due to detailed questionnaires developed and distributed by the Royal Ontario Museum in the late 1920s. Although these firsthand accounts are invaluable, data based on the memory of events that took place four or more decades earlier are apt to contain inaccuracies. In addition, the value of a recollection related to the size of a nesting is limited not only by memory but the subjective nature inherent in that kind of assessment. (These nestings are listed in the table at the end of the chapter.)

Still, these possible Ontario pigeon massings are intriguing, if not haunting and perplexing. If indeed huge, largely unmolested nestings were pumping hundreds of thousands of pigeons into the population throughout the 1870s, the virtual extinction of the species in a twenty- or thirty-year span becomes almost inexplicable. The key variable is the level at which the nestings were exploited. That most of Ontario was less economically developed and supported lower densities of human inhabitants than the U.S. states to the south meant that the intensity of killing was also probably less. But although precise information seems to be lacking, it is known that commercial operations in Bruce, Huron, Simcoe, York, Welland, and Lincoln Counties sent multitudes of live and dead pigeons to Toronto, Montreal, New York City, and especially Buffalo.[1]

The most detailed accounts on the subject address northern Bruce County, which was particularly remote and distant from rail lines. In 1872, a sizable aggregation of pigeons engaged both the white settlers and the Chippewa from Cape Croker in the "killing, storing, and vending" of the birds. Even though the number of participants was small, the quantity of birds they took "would have done credit to an army." Four years later, the April 28 issue of the *Paisley Advocate* reported that a large nesting in Annabel Township was "visited by scores of persons, and all the shot in Owen Sound and Southampton seems to have been fired away as a telegram has been received in Paisley asking for a supply."[2]

As the places suitable for nesting decreased in number and size, more birds were drawn to those areas that had the food supplies necessary to sustain them. The vast nestings in the 1870s did not reflect increases in overall population, but rather a concentration of the decreasing number of surviving birds in fewer and fewer places. Due in part to an unusual stretch of favorable weather, these years saw heavy production of beech mast in the fall of every odd year. Thus, in the spring of even years the pigeons congregated to nest in the beech forests of Michigan, Ontario, and Pennsylvania, and in odd years they sought the oaks of Wisconsin and Minnesota. It is possible that during a few of these years, neither acorns nor beech nuts were in sufficient quantity to support all of the birds. They would have been forced to nest in smaller, scattered groups or to rely on less favored food sources. In any of these events, the number of progeny that year would probably have been reduced even without the constant harassment by people. But know with confidence that wherever the birds did gather within the ken of *Homo sapiens*, there was slaughter.

THE ALLEGHENY MOUNTAINS OF PENNSYLVANIA: 1870

The year 1870 saw millions of pigeons nesting along Potato Creek in McKean and Potter Counties, Pennsylvania. Filling up the forest in a strip forty miles long and up to two miles wide, the nesting was the largest in the vicinity since 1830. Professional pigeoner J. B. Oviatt began his career with this mass of birds, situated not far from his home in Smethport. Although he had caught some pigeons as a youngster while assisting his father, the arrival of the birds in March of 1870 prompted him to buy his first net, fourteen feet by twenty feet, for a dollar. His first hauls were modest, but of seventy-five that he caught on March 22, he sold several live birds as stool pigeons for a total of $8.75, while the dead ones went for seventy-five cents a dozen.[3]

The beeches had fruited bountifully during the fall of 1869, auguring the arrival of pigeon throngs come spring. Thus tipped off, the "pigeon men" flocked to Smethport in numbers almost as thick as the birds. The local newspaper editor observed that they were fixing nets and other equipment for the "anticipated slaughter." Catches were scanty through early April, but for the next month thereafter, "a net was set in almost every clearing large enough to hold one, the whole length of the Potato Creek valley and down to the Allegheny River." The most productive operations were at the feeding areas about ten miles from the nests.[4]

Besides the pigeon men, Seneca from the Salamanca reservation routinely showed up at large nestings in the region. They dispatched scouts to identify the best place to set up their village, and then entire families, with their horses and belongings, would move in like an army. They had neither guns nor nets, focusing mostly on the squabs, which they preserved in substantial quantities. Some of the white pigeoners boarded with them, and everyone seemed to enjoy each other's company.[5]

The second stage of killing commenced several days before the squabs could fly. The pigeoners would them begin cutting down the trees that held the most nests. Eventually, the landowners, objecting to the loss of their valuable timber, forbade the practice, so the men would have to scale the trees in pursuit of the fat babies. As in other places, the Seneca used blunt-edged arrows to knock the squabs from their nests.[6]

By May 12, things in town had quieted down, as the pigeon excitement abated. The Smethport newspaper estimated that five or six tons of pigeons had been shipped from local points every day for two or three weeks. Never again would the pigeons nest in such profusion in the state.

THE LARGEST NESTING OF ALL: WISCONSIN IN 1871

The stupendous pigeon nesting that spread across 850 square miles of west-central Wisconsin in 1871 was "discovered" by A. W. Schorger as part of his

Map of the 1871 Wisconsin nesting. © Gary Antonetti/Ortelius Design, based on a map in A. W. Schorger, *The Passenger Pigeon: Its Natural History and Extinction*

decades-long project to examine the state's newspapers for information on wildlife. It was previously thought that the 1878 Petoskey nesting was the greatest of all, but Schorger found that this Wisconsin concentration "was so much larger that one hesitates to believe the evidence."[7]

Most of the birds seem to have entered the state through the Rock River and Mississippi River valleys, for few were noted along Lake Michigan or in the southwest corner. Throughout March, large flights traversed the skies over such towns as Beloit, Janesville, Lodi, Fond du Lac, Baraboo, and La Crosse. A newspaper from the last-named city said the birds "darken the vernal atmosphere," while a description of events over Fond du Lac speaks of "flocks without-any-end-either-way."[8]

The pigeons began to breed in April. The nesting area went from Kilbourn City (now called Wisconsin Dells) northwesterly to Black River Falls, then headed almost straight north to Grand Rapids (now called Wisconsin Rapids). (See map.) This huge area was not uniform, containing swamps as well as stands of mature white oak. Most of it, though, featured a surface of sand, ideal for the short, squat Hill's oak (*Quercus ellipsoidalis*), which produces a small acorn that was relished by the pigeons.[9]

After ignoring whatever pigeons appeared earlier in the spring, the *Kilbourn City Mirror* finally commented in its April 22 issue, "For the past three weeks they have been flying in countless flocks that no man can number." Even if the editor was slow in reporting the pigeons, the townspeople had mobilized a week earlier: "Had a stranger looked on to the street in town on Friday night he would have thought it about war time, or soon after an Indian scare or massacre. Young men and old, women and children, fathers, sons and husbands, and other men had a gun or wanted to borrow one. Clerks and proprietors were pouring out shot like hail in the March equinox." Later in the evening, wagons came back in from the field loaded with pigeons.[10]

The professional netters soon followed, at least six hundred of whom registered at local hotels. Area farmers and their families were recruited to further increase the ranks of pigeon catchers. It was estimated that during the nesting season over one hundred thousand people from all parts of the country visited. A few of these came just to observe, but most sought some sport or the opportunity to procure some extra income. One merchant in Sparta sold sixteen tons of shot over the season, and the same amount of powder. This amounted to about 512,000 rounds. The pigeon dealer H. T. Phillips said that his shipping operation consumed three floors of ice from a capacious icehouse.[11]

Journalist Hugh Kelly made his own inspection of the pigeon city. He and his party started from Kilbourn City and the first woods they came to consisted of trees that, without exception, were covered with nests, some having as many as thirty. But the nests were all vacant, for the pigeoners had already worked the sector. Smashed shells littered the forest floor. Occasionally they encountered dead pigeons, those that had eluded their slayers even though mortally wounded. Traveling on, the party did find throngs of healthy pigeons, enabling them to spend all afternoon and dusk "waging war against the birds." They brought back 250.[12]

It pays to cultivate the local media. The editor of the *City Mirror* thanked Frank Hills for dropping off a batch of pigeons. He continued with this endorsement: "Frank is the pioneer of the pigeon business, knows all the grounds, and has teams ready, with careful drivers, to take hunters or trappers to the pigeon grounds. His terms are reasonable."[13]

One visitor who engaged Hills to take him on a tour of the nesting was an unnamed writer from the *Fond du Lac Commonwealth*. As the journalist made his way to Kilbourn City from Fond du Lac by train, he found hunters boarding at every stop. By the time he reached his destination, there were twenty-seven of them. Kilbourn City itself was described as consisting of "innumerable coops of pigeons," evidence of the live-pigeon trade.[14]

The guide's "wholesome voice" woke the writer and his party so that they could reach the pigeons well before dawn. Having ten miles to cover, Hills forced the wagon team to roll at maximum speed. It was still dark when they arrived, so their perception of the pigeons was exclusively aural: "The indescribable cooing roar produced by uncounted millions of pigeons, as arousing from their slumbers they saluted each other and made up their foraging parties for the day, arose from every side, created an almost bewildering effect on the senses, as it was echoed and re-echoed back by the mighty rocks and ledges of the Wisconsin [River] bank."

Hills sent the men stumbling almost blindly through the brush to spots that afforded superior shooting. But the sounds made by the eruption of many millions of tom pigeons as they embarked on their morning feeding foray rattled the novices into near paralysis. It took the would-be shooters a bit to retrieve their nerve, by which time the first few flocks had passed unscathed. Then the enfilade commenced:

> Hundreds, yes thousands, dropped into the open fields below . . . The slaughter was terrible beyond any description. Our guns became so hot by rapid discharges, we were afraid to load them. Then while waiting for them to cool, lying on the damp leaves, we used, those of us who had [them], pistols, while others threw clubs, seldom if ever, failing to bring down some of the passing flocks . . . Below the scene was truly pitiable. Not less than 2,500 birds covered the ground.

When the morning flight subsided, the writer and a few others continued several miles until they reached the actual nesting place. The squabs were as plentiful "as counterfeit currency at a circus door." Most of them were not yet adept at flight and could easily be obtained. The hens also provided easy targets, the only limitation being the time it took the marksmen to load, aim, and shoot. Then there was the collateral damage that would never appear in any of the official tallies: "Many of the young pigeons were dead in their nests, the mothers probably having been killed, and her young starved."

By the end of May, the largest documented nesting of all time disbanded as many of the huge flocks headed toward Minnesota. How many birds came together in the sandy oak barrens of central Wisconsin in the spring of 1871, and how many failed to survive the onslaught? As to the first question, Schorger took great pains in defining the size of the territory occupied by the pigeons. He then multiplied that by the number of birds per acre (twenty-five trees each with five nests or ten adults). Two hundred fifty birds times 544,000 acres (850 square miles) equals a total of 136 million nesting pigeons. This would not include non-nesting adults or the squabs, most of which were probably short-lived given the thoroughness with which they were collected.[15]

To answer the second question, Schorger accepted as his basic data that a hundred barrels of pigeons (each barrel held three hundred birds) were shipped each day over the forty days of the season. This gave a total of 1.2 million dead pigeons, which he acknowledged was a conservative total. Given that the number of barrels sent out daily varied between one hundred and two hundred, the figure seems to be woefully low. And as he acknowledged, an accounting based on the number of barrels does not include birds shipped alive and birds that were never shipped at all because they were consumed locally, left to rot, or were squabs that starved.[16]

Schorger believed that "virtually" all of the pigeons that still survived in 1871 nested in Wisconsin that spring, most of which were part of the big nesting. Depending on what constitutes "virtually," he may well have been right, but at least three other large nestings were reported that year: one in Cochrane County, Ontario, which was said to involve millions of birds, and two in southeastern Minnesota, the largest and best documented of which occurred near Wabasha. A contemporary newspaper dispatch stated that the pigeon city at Wabasha was thirty miles long, though a resident many decades later said it was only seven miles long by a half mile wide. But even at only two thousand acres, a colony where every oak had multiple nests likely held several million birds.[17]

Schorger may have glossed over the Wabasha nesting, but the hunters did not. The *Wabasha Herald* of May 11, 1871, provided this profile of local pigeon happenings:

> There is an electric something about these clouds of birds as they wheel and circle about here in countless myriads that sets every sportsman's blood bounding. All the guns in the country have been brought to bear on the game . . . Powder and shot is a scarcity in the market, butchers are thinking of suspension, dry goods

> clerks have taken unceremonious leave of absence for indefinite periods. Immense numbers are slaughtered. Large quantities, alive and dressed, are shipped daily to market.[18]

SHELBY, MICHIGAN: 1874 AND 1876

Little has been recorded about the large nesting at South Haven, Michigan, in Van Buren County in 1872. One old pigeoner recounted that they had located their operations in Bangor, ten miles to the southeast, and caught many birds during some heavy snowstorms. A local paper reported at the time that over a forty-day period a total of 7.2 million birds were shipped out in barrels by the railroad. No mention was made of the trade in live pigeons.[19]

The major nestings of 1874 and 1876 formed in roughly the same place: Shelby, Michigan in Oceana County. Perhaps nowhere else did the pigeons create a more profound effect. Some said their appearances were acts of Providence.

The principal city in Oceana County was Pentwater. Situated on Lake Michigan, it was endowed with streams, a superb harbor leading to a small lake, and two branches of the Pentwater River, which provided easy access to inland timber and farming regions. Another up-and-coming municipality was Hart, founded on the banks of the south branch of the Pentwater River. The river provided the power for the county's first gristmill, which anchored the town's economic growth and probably led to its selection as county seat. Bereft of these advantages, Shelby seems to have been founded as the halfway point between Pentwater to the north and Whitehall to the south. It was a place where travelers could spend the night. Later, an artificial tributary in the way of railroad tracks was completed. Being connected to the larger world boosted morale and commerce, and growing prosperity seemed inevitable. But the novelty of the train wore off, and the surrounding farmland, though promising, was too hilly to be fully utilized. An 1890 history says that "a period of decline was becoming painfully manifest . . . as the little village was sinking into the slough of despond."[20]

Then, in the early spring of 1874, the pigeons arrived. They congregated near the Lake Michigan shoreline, in woods of hemlock and pine twenty miles long and from four to seven miles wide. Nesting began in early April, and the first chicks hatched after another two weeks. As the males and females took their turns pursuing beechnuts and worms that might be twenty-five miles away, the birds passed without break over some locations for hours at a time. Here hunters would gather to shoot pigeons until they became bored or dusk intervened. Those who spent the entire day routinely bagged 250 to 300 pigeons.[21]

Shelby hotelkeepers and the railroad organized a weekend excursion for hunters. A special train originating in Chicago arrived on one Saturday night with a hundred pigeon hunters from various places along the line. One observer noted, "During the whole day, in and out of the roost, it was the most lively fusillade I ever heard. They set a large belt of woods on fire in the roost, and if it had not been for an opportune rain, great damage would have been done."[22]

All the locals became pigeon hunters or agents, and they were joined by six hundred professional netters. One fortunate netter took 154 dozen pigeons in a day, while the record catch for a single haul of a double net was 140 dozen. It was estimated that on one day at the height of the nesting, forty-two thousand pigeons were either discarded due to spoilage or shipped from Shelby. Over four weeks, the same accounting arrived at an average of twenty-five thousand shipped per day. Another source, a dealer working the nesting, said he personally shipped 175,000 pigeons out of Shelby and that over the thirty days when activities peaked 900,000 pigeons left Shelby in barrels. Those catches were the heaviest he had ever seen in his decades in the business. Because of the abundance of the pigeons, prices at Shelby were low: dead birds brought thirty cents a dozen and live ones brought to the depot went for fifty cents a dozen. When all was said and done, the people of Shelby proclaimed the pigeons an elixir, for their slaughter pumped $50,000 into local coffers. By their deaths, the birds had breathed new vitality into the foundering village.[23]

In 1876 an even larger mass of the feathered manna nested near Shelby. This time they took up an area twelve miles long and three miles wide. The nesting site was described as "deep evergreen timber," and some of the trees were so encumbered with nests they bent to the ground. Given the noise and bustle of the pigeon colony, resident A. S. Souter said, "A person had to speak at the top of his voice to be heard a few feet. The birds seemed to have no fear and the squabs were too young to know that man was their enemy." Adults on their feeding flights would darken the sky from early morning to late afternoon.[24]

The nesting and Shelby both received national attention. Clearings and wet areas amid the hills were soon occupied by the five hundred professional netters who descended upon the town. "These locations netted their owners a good sum in rentals," remembered Souter. "I rented my farm to many hunters, making a charge of ten dollars a hunter." One estimate claimed that over seven hundred thousand birds were taken: 1,781 barrels, 1,928 coops of live pigeons, and 2,000 dozen kept for feeding and later slaughter.[25]

Edward T. Martin, a major game dealer from Chicago, would write two articles, one in 1878 and another in 1914, defending the pigeoners and absolving them of any blame in the extinction of the pigeons. Comparing Shelby to Petoskey, Martin wrote that at the former place, "the birds were more 'come-at-able,' easier caught, easier shipped." Live birds were brought to pens in such volume, there was insufficient food and water to support them, and "half had fretted themselves to death, or else perished for want of food and drink." The glut of pigeons on the Chicago market depressed prices so much that a barrel of pigeons could not even bring fifty cents. Surplus barrels were simply discarded as garbage. In Martin's view, the huge number of birds that appeared at Petoskey two years later proved that nothing done at Shelby appreciably diminished the pigeon population.[26]

The blood of the pigeons did not go unappreciated. It was, in the words of the old history, "the golden shower thus poured upon the village" that allowed Shelby to invest in various improvements. Farming expanded and the population grew. Soon, Shelby was every bit the equal to its Oceana neighbors in stature and economic vitality.[27]

WARREN COUNTY, PENNSYLVANIA, AND PETOSKEY, MICHIGAN: 1878

Although well documented by the local newspapers, the large Pennsylvania nesting in Forest and Warren Counties has largely been ignored in the extensive literature of the species in this state. A pity, too, for the carnage at this late date was spectacular. Perhaps it was merely overshadowed by Petoskey.

The birds first appeared over Warren on March 7, flying at heights beyond the reach of firearms. But over the next two weeks, they settled in to nest, and the serious killing began: "Last week nearly 100 barrels of dead birds were shipped from Sheffield. At this rate the pigeon will soon be exterminated." But, in fact, the numbers of dead pigeons just swelled.[28]

Shooters and spectators made it difficult for the netters to do their work, for they tended to move in too close and scare the birds. But some of the professionals persevered and were rewarded. "The gang near Balltown," for example, "are scooping them in at a great rate. They took 80 dozen in 2 days." They sent the birds to New York via Sheffield and received $3 a dozen. By April 20, the freight agent at Sheffield had tallied 291,741 birds shipped from that depot. It was estimated that nearly the same number left Tionesta, and forty thousand from Tidioute. For this total of over six hundred thousand birds, $75,000 was received.

For the next month and a half, the killing continued, with some shifting of locations as areas became denuded due to the hunters' efficiency or the pigeons' flight. On May 28, one of the papers reported, "The trappers are now operating near Brookston and also about Kane and altogether the pigeons get no rest at all." Pigeoners were still taking birds when the final figures were announced on June 11: over seven hundred thousand pigeons had been shipped from Sheffield and two hundred thousand from Kane. That did not include the ninety thousand taken earlier at Tionesta and Tidioute, nor those "carried away by shootists" and the two thousand dozen still in coops awaiting transport.

But no one ignored the Petoskey nesting. It was, by far, the best documented of any, and it included a full cast of articulate heroes and villains. About forty miles in length and three to ten in width, the nesting of 1878 occupied large chunks of three counties at the north end of Michigan's lower peninsula: Emmet, Cheboygan, and Charlevoix. Each of the county seats had a newspaper at the time, which makes it possible to reconstruct what it was like that spring, when the region would be invaded by the pigeons and the throngs of people who followed after. It was as if oil had been discovered, for the birds were the center of an industry, and their presence turned the region into a boomtown complete with hundreds of pigeoners from all over the country; pigeon dealers and agents; hordes of nearby Indians looking for work; pluckers, shuckers, pickers, and packers; clerks to keep track; and there is even one mention in a secondary source of "trollops." But unlike any other place consumed by pigeon fever, there appeared as well a small group of men who tried to stop the killing. One wrote of the experience, which prompted both rebuttals by detractors and corroborations by supporters. That, too, was unique.[29]

Most residents of the Upper Midwest examine early March closely for even the faintest signs that winter is giving way to spring. But the winter of 1877–78 had been so mild that one newspaper editor commented that there could be no bursting of spring because there had never been a winter. Typically, winter mail deliveries required carriers to cross the frozen Mackinac Straits by foot, but since March 2 the ice had been too thin. Several careless people who'd ventured onto the thinning ice had had to be rescued, and two had drowned in the process. To prepare for summer, the firm of Dingman

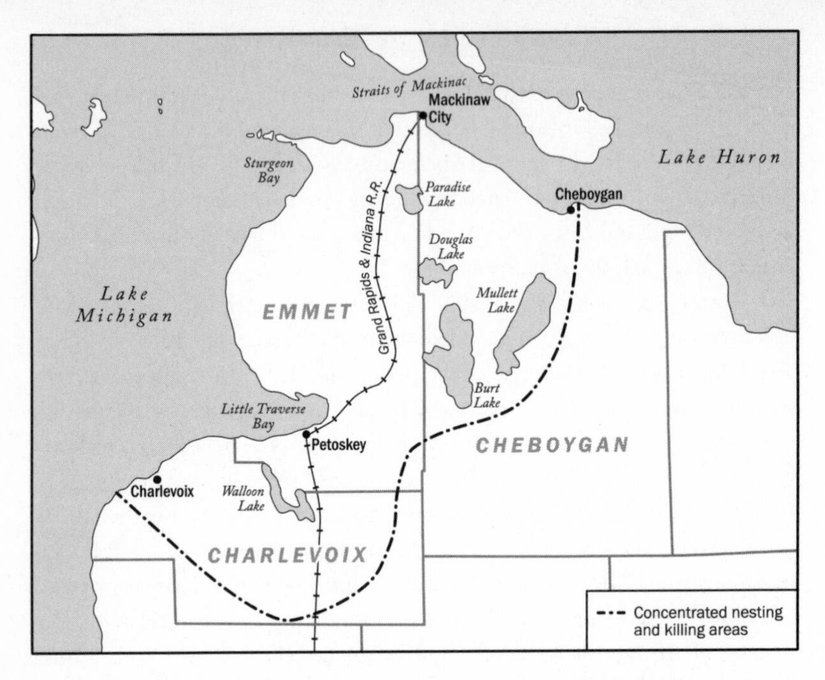

Map of the 1878 Petosky, Michigan, nesting. © Gary Antonetti/Ortelius Design, based on a map in Reginald Sharkey, *The Blue Meteor*

and Franks began filling their icehouse by towing the drifting ice to shore for collection, even though it was light and of poor quality.[30]

Inland the warmth had created other problems. Timber operators hauled their logs to mills over frozen ground. But the unseasonable weather meant that they had but the early-morning hours to work before the hard surface of the primitive roads thawed into deep mud. These gooey conditions resulted in a significant reduction in product. Later in the spring, around Petoskey, fires broke out in several places. Charles Bemis was hauling a load of hay that ignited, and he was badly burned as he fought the flames in a successful effort to save the silage.[31]

A run of herring provided excellent sport and fodder for anglers using hooks and lines from the Charlevoix docks. Robins appeared on March 9. Six days later, Rozelle Rose, editor of Petoskey's *Emmet County Democrat*, noted the arrival of the season's first passenger pigeons. Several large flocks had flown over, prompting him to query, "Where are the sportsmen with their guns?" He did not have to wait long for an answer.

Within a couple of weeks, the woods became filled with swarming pigeons and the gun-toting legions of men and boys who pursued them. One party of three hunters had just bagged four hundred pigeons, an easy task given that "great flocks were seen flying in all directions." But these three were not alone, as throughout the area hunters were racking up totals just as high.[32]

The actual nestings occurred north of Petoskey in low woods of pine, maple, and beech. There were seven or eight places where the birds concentrated. The first big flocks took up residence at Pickerel Lake, while birds arriving later found suitable habitat at Cross Village, Burt Lake, Crooked Lake, and in the swampy woods along the Maple, Crooked, and Indian Rivers. Toward the end of May, masses of pigeons began collecting in what was probably an unsuccessful nesting effort at Boyne Falls to the south. Pigeon trappers filled the town's hotel, but by early June most had returned north, as pigeons were difficult to find.

By April and May, pigeon mania had infected everyone. In Petoskey, McCormick and Brothers, a firm specializing in frozen fish, diversified, sending daily shipments of pigeons to Philadelphia, New York, and other large cities. Hotels and boardinghouses ran out of space, and porters were dispatched to comb the town for places to accommodate luckless guests. W. T. Latham, the town's dairyman, reported that business was unusually brisk. Any boy who wanted could find a job plucking pigeons. In Cheboygan, double-barrel shotguns were at a premium, and supplies of ammunition were nearly exhausted. But the pigeons themselves were in such abundance they became "a drug upon the market," forcing dealers and merchants to give them away for free.[33]

Mark Craw, who was eleven at the time, provided a firsthand account of what it was to be a youngster in the midst of the great nesting. He was too young to possess a gun, but he acted as an errand boy. When the shooters found themselves running low on bullets, they would dispatch Craw to the Hannah and Lay store in downtown Traverse City to get more. He would also serve as a retriever, collecting dead birds and adding them to the growing pile. Eventually the corpses would be divided up among the hunters, with a few going to Craw for his efforts. The site from which the pigeoners plied their trade eventually became the Traverse City State Hospital.[34]

Among the pigeoners were some characters whose strangeness rose above the level of their colleagues' and granted them attention. There was, for instance, Jackson Bennet, who walked all the way from his home in Shelby to Petoskey, 191 miles. He spent a month trapping and obtained

151 dozen pigeons. Toward the end of his stay he caught a black bear, which he had with him as he started his long trek homeward. Old Joe, a local whose surname went unrecorded, had lost an arm in the Civil War. But his handicap did not slow him down any as he helped work the nets. With one motion, he would grab a pigeon by the leg and toss it into his mouth head-first, then chomp down on the skull: "What a sight! His face was smeared with blood from ear to ear; his beard dripped gore; and his clothes were covered with it."[35]

But the pigeons attracted more than pigeoners to Petoskey. At this time in Michigan, the only effective voice for wildlife conservation was the Michigan Sportsmen Association (MSA). This was a coalition of sportsmen clubs that had formally come together three years earlier. The state had had various sportsmen clubs for a long time, but they'd started as primarily social gatherings. The passage of years, however, made it clear that the objects of their searches were fast disappearing, so some of them began actively fighting for legislation to protect wildlife. More and more of them also broadened their views to appreciate how important game was as food for rural citizens and the critical role many birds played in the control of insects. Some even adopted the modern notion that the native flora and fauna were valuable for their own sake. According to the president of the MSA in an 1880 statement, of the three such state associations in the country, the Michigan organization was the only one "whose main object is not the slaughter of the pigeons [in shooting matches], but the protection of its furred, finned, and feathered game."[36]

Taking their missions to heart, the Bay City and East Saginaw MSA chapters met in early April to discuss what could be done to stem the pigeon killing at Petoskey. Most galling was what appeared to be the total failure to abide by or enforce the two modest restrictions that Michigan law then imposed on the taking of passenger pigeons. The first was to limit the use of firearms to no closer than five miles from a nesting. Netters, on the other hand, could operate as close to the nesting as two miles "at any time from the beginning of the nesting until after the last hatching of such nesting." Conviction for violating these statutes could bring a $50 fine plus costs. It must be noted that once the eggs hatched, the netters had a free hand in taking as many adults or squabs as they could.[37]

The clubs decided to do something that had never before been done anywhere in passenger pigeon range: they were going to send a contingent to the Petoskey nesting to enforce the pigeon law in an effort to save birds. The Bay City chapter secured the services of undersheriff William Fox, and East Saginaw hired R. Fairchild, both described as "two thorough sportsmen and well chosen for the work in hand." Accompanying then as "backers" were Del McLean and Henry B. Roney.[38]

This unique intervention came about due to the influence of two remarkable men, William Butts Mershon of the East Saginaw chapter, and Roney, also of East Saginaw and secretary of the MSA. Mershon was a lumber baron who loved to hunt and fish. He owned a railcar, which he took all over the Midwest in pursuit of game. But at some point he realized that the quarry he so enthusiastically sought was vanishing. He became a leader in promoting conservation legislation in Michigan and on the federal level, using his influence and fortune to get things accomplished. He attained immortality in 1907 by producing the first book-length account of the passenger pigeon.

Henry Roney, in his way, was even more intriguing. A music teacher born in Ohio, he came to Saginaw in 1871 as organist for the First Congregational Church United Church of Christ. Later he would gain international fame with his traveling choir known as Roney's Boys. But while in Saginaw, he worked closely with Mershon in a successful effort to get the state legislature to pass a game-protection bill.

On April 15, Fox, Fairchild, McLean, and Roney boarded a train for Petoskey. They secured lodging at the Cushman House hotel and encountered "Uncle Len" Jewell of Bay City, a savvy old woodsman who agreed to join them for a few days. A check of the hotel register revealed that guests from at least twelve states were present, including those as far away as New York, Virginia, and Texas. Roney noted that "in the village nothing else seemed to be thought of but pigeons." Fifty wagons were in constant motion taking pigeons to the railroad station, and the road upon which the wheels rolled was covered with pigeon feathers: "The wings and feathers from the packing-houses were used by the wagon load to fill up the mud holes in the road for miles out of town."[39]

Wasting no time, they hired a wagon and went in search of the nesting grounds. Fifteen miles out of town they came to a track heading into the woods toward the cacophony of pigeon noise that told them they were close. Roney and two others went on foot, following behind the wagon. They did not have far to go:

> The twittering grew louder and louder, the birds more numerous, and in a few minutes we were in the midst of that marvel of the forest and nature's wonderland—the pigeon nesting. We stood and gazed in bewilderment upon the scene around and above us.

Picture of the rail car owned by William Butts Mershon that he used on hunting trips during the 1880s and '90s. This one displays assorted game shot by his party in North Dakota. Courtesy of Local History and Genealogy Collection, Public Libraries of Saginaw (Michigan)

Henry Roney and his choir, circa 1900

> Was it indeed a fairy land we stood upon, or did our eyes deceive us? On every hand the eye would meet these graceful creatures of the forest, which, in their delicate robes of blue, purple, and brown, darted hither and thither . . . In every direction, crossing and re-crossing, the flying birds drew a network before the dizzy eyes of the beholder, until he fain would close his eyes to shut out the scene.[40]

While entranced by the sight of the birds, they had not come as naturalists or journalists, so they proceeded to where the pigeoners were most active. Bedlam reigned as an army of men and boys felled the nest-laden trees to get at the squabs. Others used long poles to knock the birds from their elevated perches. Upon grabbing their quarry, the pigeoners would yank their heads off. Almost an equal number of squabs lay dead among the understory, having starved when their parents failed to survive the gauntlet of nets and guns. The going rate for this work was a penny a bird, and every able-bodied male in the vicinity was so employed. Roney and his crew, though, wanted bigger pigeoners to pluck and obtained the names of the leaders. With that information in hand, three of the party headed back to Petoskey.[41]

Roney and Jewell, meanwhile, continued their reconnaissance by walking five miles north to Maple River. They were intent on busting illegal netting operations but found themselves stymied as no one would acknowledge that there was any netting. They met two pigeoners who were willing to chat a little, but became close-lipped as to where the netting was occurring. A "tow-headed mossback" insisted he knew nothing of pigeons, yet feathers littered his residence. Similarly, "an old hag" known to be lodging three netters swore that "there wasn't no nesting anywhere around here that she knew of," even as she stood at her stove frying pigeons. When they reached a ramshackle hotel, the two men adopted a stratagem that proved fruitful. Jewell left the hotel while Roney stretched out on a bench and feigned sleep, "emitting snores that would have done credit to a full grown porpoise." Convinced by Roney's cetacean impersonation, the pigeoners ignored him and began talking freely, thereby providing him with the intelligence he sought. Perhaps even more helpful, though, was when Roney and Jewell began cultivating children and found one youngster who gave quite detailed information on the closest netting operation.[42]

Roney and Jewell began their trek to the netting area early the next day. After a long hike they reached a ravine broad enough to hold the bough house and net. They stationed themselves behind the shelter of some bushes to await the crime they were certain would soon occur. The pigeons began alighting in the mud seeded with salt and sulfur until they became a seething mass of blue and purple. Then, with a loud whoosh, the double net of ten by twenty feet slapped over them, trapping hundreds of birds. The two detectives leaped from their hiding place and beheld a disgusting sight:

> In the midst of them stood a stalwart pigeoner up to his knees in the mire and bespattered with mud and blood from head to foot. Passing from bird to bird, with a pair of blacksmith's pincers he gave the neck of each a cruel grip with his remorseless weapon, causing the blood to burst from the eyes and trickle down the beak of the helpless captive . . . When all were dead, the net was raised, many still clinging to its meshes with beak and claws in their death grip and were shaken off.[43]

The netter said he had been in this line for years and had caught as many as eighty-seven dozen in a day. Roney later learned that on that very day the netter had come close to his record with eighty-two dozen. As the net was within a hundred rods of the nests, Roney and Jewell graciously said good-bye and hightailed it to Petoskey to swear out a warrant. Faced with two eyewitnesses testifying against him, the netter pleaded guilty and paid the $50 fine.[44]

More than once, Roney and his group were threatened with being "buckshotted," but perhaps surprisingly they experienced no actual violence. The closest they came to suffering vengeance is when Jewell, feeling "the rheumatics," decided to go fishing one day rather than track down offenders. He brought back a dozen speckled trout, which he proudly displayed. He gave half of them away and asked the hotel to prepare the others as dinner for the group. Unfortunately, he had forgotten that trout season did not start until May 1, and he was immediately arrested and fined $25. Roney says they were all shocked by Jewell's actions, and he assured the officials that there were no hard feelings, for in arresting Jewell "they had done no less than their duty."[45]

Roney's raiders also received strong admonishment that they go after out-of-town big shots and not just target locals trying to eke out a living. The most conspicuous of these visiting transgressors was the editor of the *Grand Rapids Eagle* and a prominent member of the Michigan Sportsmen Association, A. B. Turner. He not only allegedly shot birds within two miles of the nesting, he did so merely to amass a high number, making no effort to collect the fallen birds. One witness was the postmaster and justice of the peace of the township, whose home was surrounded by the woods where the pigeons nested. After watching Turner and his party fire at the passing flocks, the man chewed them out for ignoring the law. Turner's group responded that unless the man backed down, their guns might accidentally go off in his direction.[46]

Numerous observers came forward to support the claims against Turner. But the prosecuting attorney was away on business, so the people's case was handled by a less experienced substitute. Representing Turner was one of the most successful lawyers in town. The dismissal of all charges at trial confirmed Roney's fear that convicting pigeon miscreants would be almost impossible in local courts.[47]

The Turner case was certainly the most notorious, but other people were tried as well. Quite a few pigeon hunters were fined $50 plus court costs. Later in April, two men from the Detroit Sportsmen Club were also arrested for shooting too close to the nesting and were fined $110. Their request to remain anonymous, so their club would not learn of their indiscretion, was honored.[48]

The local papers were sympathetic to Roney's activities. Editor Rose warned that more "squab detectives" were arriving every day and that violators would surely be caught if they did not obey the law. Unfortunately, once the four men from Saginaw and Bay City returned home, there were no others to replace them. But the specter of fear created by Rose's imaginary enforcers might have scared off a few wannabe pigeon pirates.[49]

How many birds were actually killed that spring and summer? Roney concluded that over 1.5 million dead birds and 80,352 live birds were sent out by rail. Only estimates could be made for boat cargoes, wagonloads, express packages, and the vast quantity of squabs and adults that died but were never secured, particularly squabs that starved to death. In adding all these together, Roney says, "We have at the lowest possible estimate not less than a billion pigeons sacrificed to Mammon." Pigeon dealer Martin provides his own detailed figures and comes up with a total of 1,107,866 birds entering the marketplace. But yet another pigeon merchant said his firm alone processed almost twice as many live birds as Martin's total.[50]

Roney's estimate that a billion pigeons died proved troubling to subsequent writers, even those sympathetic to conservation. Some quoted Roney's detailed figures, but ignored the concluding estimate, while others specifically stated they thought it is an exaggeration. Martin not only mocks the number, but claims that the birds nested three times that year, and that some of the nestings were in places so remote they escaped all exploitation. (No one brought up the Pennsylvania carnage going on at the same time.) It matters little whose estimate was more accurate, for we do know with absolute certainty that with the conclusion of the 1878 nesting season, the passenger pigeon would never gather in such numbers again.

BENZIE COUNTY, MICHIGAN: 1880

There were evidently two significant nestings in 1880. The lesser known of these occurred in western Pennsylvania, where it was averred that nine tons of pigeons originating from McKean, Forest, and Warren Counties passed through Emporium by express on April 1. Another nesting area in Potter County was a mile long and half mile wide, but impatient shooters caused the birds to scatter without fledging any young. In June, yet another nesting was attempted in the county at the head of West Branch. The shooting was reported as good, and "a great many" pigeons were netted on salt beds.[51]

Much more attention was devoted to Michigan. Large flocks also returned to the Petoskey region, but upon discovering frozen lakes and deep snow, they retreated south to Benzie County. They chose an area of woods straddling the Platte River that ran ten miles long and up to four miles wide. A local writer said of their arrival, "They came in clouds, millions upon millions. It seemed as if the entire world of pigeons were concentrating at this point. The air was full of them and sun shut out of sight, and still they came, millions upon millions more."[52]

Telegraphs and railroads made it possible for a growing number of people to actually visit large nestings. In two weeks time, as word of the pigeon gathering spread, Benzie County was inundated with three thousand hunters, described as running the gamut: "Professionals, amateurs, mossbacks, city sports, young bloods, and greenhorns had invaded the country from all directions, surrounding and penetrating the pigeon grounds." Nearly every residence with a spare bed became a boardinghouse offering accommodations to the visitors.[53]

Except for the speed with which word of the nesting traveled, most of this sounds pretty familiar. But there was something new. The old hunters noticed that these birds behaved differently from usual. A core of mated birds commenced nesting, but they were outnumbered ten to one by other birds that seemed reluctant to get on with business. Less patient than the pigeons, bands of hunters entered the nesting area and started blasting away at birds and nests alike. Under this regimen of harassment, combined with a storm that brought eight inches of snow and vigorous winds, "many of the roosting birds, disgusted, postponed their anticipated housekeeping and scattered."[54]

Many birds did stay, however, although after the first huge group of pigeons moved on, others may have followed some days later. One reporter who tried to see them encountered locals who had perhaps learned something from the Petoskey nesting. He was without gun or net and so was suspect in the eyes of those he asked for directions. Men gave bogus instructions while women and children feigned ignorance. What had been purported to be but a three-mile hike to the nests turned into one of fifteen. But unlike at Petoskey, where at least a few violators were arrested, no law enforcement occurred at this nesting, and amateur hunters shot nests and their inhabitants without pause.[55]

Some of the pros were disappointed with the season, as many of the birds departed without nesting. But others did quite well. Dr. Isaac Voorheis of Frankfort, Michigan, caught 1,316 birds with a single snap of his net. Six snaps netted him $650 worth of pigeons.[56]

Simon Pokagon visited the nesting and was filled with shame and pity. Many of the nests were placed in mature white birch, the bark of which peels like flammable paper. With the birds scattered and many disappointed pigeoners around, Pokagon witnessed two techniques for procuring pigeons that he had not seen before: "These outlaws to all moral sense would touch a lighted match to the bark of the tree at the base, when with a flash more like an explosion, the blast would reach every limb of the tree, and while the

Petoskey memorial, placed in 1957 to commemorate the 1878 Petoskey nesting, stands at the Michigan State Fish Hatchery at Oden, Michigan. Photo by author

affrighted young birds would leap simultaneously to the ground, the parent birds, with plumage scorched, would rise high in the air amid flame and smoke." The fat squabs, so rudely dispossessed, would often splat upon hitting the ground. Pigeoners collected thousands of fledgling that way over the day.[57]

That evening Pokagon stayed with an old trapper who was camped just outside the nesting ground. Pokagon explained his anguish at what he had witnessed earlier in the day. The trapper appeared moved and promised to show him the way he caught birds, a method "that will please any red man and the birds too." Early the next morning they headed to the trap site not far away. The trapper threw handfuls of corn as enticement for the pigeons. After a short wait, the birds descended on the bait and fed voraciously. As Pokagon admired the sight, he noticed the pigeons beginning to topple, beating their wings to no effect as they lay on their sides quivering. The trapper leaped from his hiding place and with Pokagon's help soon caged a hundred birds. Pokagon thought to himself, "Certainly Pokagon is dreaming, or this long-haired white man is a witch." The old man revealed his secret: the corn had been soaked in whiskey. Pokagon left, feeling both betrayed, as the trapper had expressed support of temperance, and overwhelmed by sorrow: "Surely the time is now fulfilled, when false prophets shall show signs and wonders to seduce, if it were possible, even the elect."[58]

POTTAWATOMIE COUNTY, OKLAHOMA: 1881

The Indian Territories, now known as Oklahoma, began to emerge as a place that harbored some of the last gatherings of pigeons. The first rail line to cross the territory was completed in 1872, but another fifteen years passed before any new tracks were laid. The limited transportation opportunities made it poor pickings for the pigeoners as long as the birds could still be taken in areas with better infrastructure. Most of the pigeons used this region as wintering grounds, including one large roost on the Baron Fork River that was utilized from at least the Civil War years to 1873.[59]

The massive nesting on the Pottawatomie Reservation (Pottawatomie County) west of Atoka in the spring of 1881 was unprecedented. Not a great deal is known about this occurrence, but according to the available information it comprised millions of birds. They took up residence in a tract of masting post oak twenty miles long and fifteen wide. A seasoned pigeoner named W. P. Thomas said he penetrated ten miles into the nesting and saw no end in sight. Pigeons covered every tree in such numbers the branches sagged.[60]

Due to its isolated location, the nesting drew only professionals and locals from the reservation. There were none of the conflicts with gun enthusiasts that plagued the netters at other locations. Thomas said that the Pottawatomie left the pigeons alone as food, preferring to hunt white-tailed deer and turkeys, both of which were readily available. Tribal members showed friendliness toward the outsiders, and many Indians found work collecting squabs for a St. Louis firm. Those birds were packed in barrels of ice and sent to cities as far away as Boston. Despite the obstacles, that St. Louis outfit Judy and Company reaped a profit of $20,000.

But Thomas specialized in live pigeons for the shooting trade. Because acorns were so abundant and the nearby Canadian River provided ample water, the usual methods of baiting proved ineffective. Most of the netting activities, therefore, took place on the gravel banks of the river, where the pigeons sought the stones they needed for digestion. There are many descriptions of netters and how they killed newly caught birds, but Thomas provides one of the best accounts of those who sought live pigeons: "The crates in which the birds are put when caught are simply large flat coops. The netters are spread over an area of twelve or fourteen miles. Every evening the teams take a round and collect all the crates. It is now necessary to get the birds 'on their feed,' or else they will die." The pigeons were then transferred to large holding pens with ample food and water. Typically, several days would have to elapse before they relaxed enough to be again crated for the next leg of their journey. Loaded on wagons, they began the arduous trip to the railroad station, where yet another holding coop awaited them before their being sent east on the train.

Thomas already had contracts for forty thousand live birds to supply the New York State Sportsmen's Association's annual shooting meet. But he lamented that the business would be much less profitable than usual due to the 110 miles and rough terrain that separated the roost from the closest railroad station: "At one time I had fifteen wagons on the road. There are several streams to be forded and the Arbuckle Mountains have to be crossed . . . It took a wagon three days to make the trip."

Lest any of his readers be concerned, Thomas ends his comments with comforting news: "Pigeons nest four times a season . . . The increase is larger [this year] than ever before. The number of squabs killed and pigeons netted is insignificant in comparison with the numbers hatched out. There are millions of them."

KETTLE CREEK, PENNSYLVANIA, AND WISCONSIN: 1882

There were two more attempts at large nestings. One occurred on Kettle Creek near Emporium, Pennsylvania. Large masses of pigeons settled in by March 23. The first of the pigeoners appeared a week later, and by the end of April they filled the local hotels. Live pigeons in astonishing numbers were brought to nearby Coudersport. A newspaper report from June 22 says, "Pigeoners still besiege the nesting and many thousands of birds, both dead and alive, have departed as freight from Potter County." The total take was estimated as between seventy-five thousand and one hundred thousand birds. And finally there is a curious comment on August 10 that pigeon corpses were accumulating along the river near town.[61]

Probably involving a greater quantity of pigeons were the well-documented nestings that occurred in several places in Wisconsin. Monroe County hosted one that was three and a half miles by four miles, and another that was nine miles long and one and a half miles wide. Breeding pigeons in Adams County occupied a section nine miles long and one and a half miles wide. The birds were drawn by the heavy acorn crop from black oaks, although many of the birds that were killed had been feeding on various domestic grains as well.[62]

As a boy the Reverend E. C. Dixon's father took him, his two brothers, and several other helpers into the pigeon city in Adams County. It was a Saturday and the party went from tree to tree with a tall ladder so they could collect the squabs from their nests. At the end of the day they had amassed eighty-one dozen fledglings for sale, not to mention another several dozen they had forced from their nests but could not salvage.[63]

Amateurs such as Dixon and his comrades did their part in reducing the nesting birds, but the professionals went the extra mile to make 1882 so devastating. Although it was unspoken, one gets the feeling that in their guts they knew the end was near, so they were committed to taking every last pigeon. In Adams County, before the birds even began nesting, one netter caught over six hundred dozen in a single day. For this, he made $600, although some of it went to helpers and the cost of shipping. More pernicious was the unnamed C, a former member of the Wisconsin legislature who credited himself as responsible for the passing of a law protecting pigeons on their nests. (Schorger identified him as W. H. H. Cash.) He hired two hundred Ho Chunk tribesmen to systematically go from nesting area to nesting area collecting squabs: "In order to get more of them and with less difficulty, the Indians have been and are still employed to shake the trees." In this way not only the young are killed but the eggs are broken. In two days this operation reportedly reaped 27,060 squabs.[64]

Adding to the carnage, other Ho Chunks, working freelance, also arrived to partake of the bounty. The women used long poles to get at the squabs; when the birds were beyond their reach, archers employed blunt arrows to knock them to earth. Some of these were kept for personal use as the old men and women devoted their time to removing feathers and drying the morsel that remained. Tribal members sold the excess squabs in great enough quantities to earn up to $13 a day. The younger the squabs, the more tender they were. Unfortunately, such tenderness did not hold up well when the product was jammed tightly into barrels with ice. To ensure that the barrels were filled to maximum capacity, the contents would "be forced in by treading and stomping." Not surprisingly, eastern dealers were less than happy upon receiving barrels full of squab pâté. This led to complaints and promises by the game associations that they would not allow this profligate destruction to be repeated in future years. And, truth to tell, it was not repeated.[65]

The *Milwaukee Evening Wisconsin* reported that one dealer received so many more pigeons than he could handle, thirteen hundred dozen spoiled and had to be "buried in a heap." In another incident, several tons of unsold birds were dumped into the Wisconsin River. This comprehensive slaughter led citizens in New Lisbon to send a petition to the governor asking that he deploy the National Guard to protect the nestings. Because the 1882 nestings were scattered, it was even more difficult than usual to get a handle on the total number of birds killed. One estimate placed the take at 2,138,400. Schorger made the point using different words: in 1850 those numbers would have been mere drops in a most capacious bucket, but in 1882 they were a large part of a pool that had substantially shrunk.[66]

From 1870 to 1882 was a critical period in the history of the passenger pigeon. What happened in those years made the bird's extinction inevitable. In 1871, the largest nesting on record occurred, but 1878 saw the final attempt of the birds to breed in vast numbers. Four years after that, the last nesting attempts of any size numbered at most a few or so million birds. By then, there were probably not enough living passenger pigeons to have kept the species going for long even if they had been allowed to breed. But, if anything, these remnants encountered even more ferocious exploitation, and so their efforts to reproduce became virtually futile.

I can't help but think that if just one large mass of nesting birds had been allowed to breed in peace, their cohort of young might have given the species an extra decade or so of life, just enough time for the species to be rescued by new laws and attitudes. But neither an organized industry nor those practicing behaviors etched over lifetimes were to be denied by a dawning reality new to their experience. And so it goes.

LAST NESTING AREAS

1870	Potato Creek, McKean County, Pennsylvania
	Mettagami River, Cochrane County, Ontario
	Kepple, Grey County, Ontario
	Ashfield, Huron County, Ontario
	Goderich, Huron County, Ontario
1871	Kilbourn City (Wisconsin Dells)/Sparta, Wisconsin
	Wabasha, Minnesota
	Chatfield, Minnesota
	Mattagami River, Cochrane County, Ontario

1872	Ulster County, New York
	South Haven, Van Buren County, Michigan
	Shawano County, Wisconsin
	Kepple, Grey County, Ontario
	Bruce County, Ontario
1873	Rochester, Minnesota
	Grey County, Ontario
1874	Shelby, Michigan
	Collingwood County, Ontario
	Welland County, Ontario
1875	Chatfield, Minnesota
	Kincardine, Bruce County, Ontario
1876	Shelby, Michigan
	Amabel, Bruce County, Ontario
	Kincardine, Bruce County, Ontario
	Abitibi River, Cochrane County, Ontario
1877	Faribault, Minnesota
	St. Vincent, Grey County, Ontario
1878	Petoskey, Michigan
	Warren County, Pennsylvania
	Abitibi River, Cochrane County, Ontario
1879	Corry, Pennsylvania
	Parry Sound, Ontario
1880	Warren and McKean Counties, Pennsylvania
	Benzie County, Michigan
1881	Pottawatomie County, Oklahoma
1882	Kilbourn City/Sparta, Wisconsin
	Emporium, Pennsylvania

Table generated by author

CHAPTER 8

Flights to the Finish

For every pigeon that was shot and recorded during the last part of the nineteenth century, probably 100, perhaps a thousand, were shot and eaten. Who was there to record them?

—EDWARD HOWE FORBUSH, 1927

THE LAST WILD FLOCKS: 1883–89

Between 1882 and 1886, there were several mass gatherings, but they were tiny in comparison to what had been but a few years earlier. Because they were small and were of short duration due to exploitation, not much is known about them. For example, a roost formed near Huntsville, Texas, in 1883. Hunters marauded the site every evening during November and December.[1]

Missouri seemed to draw most of the birds and attention during the next two years. W. W. Judy, the pigeon dealer in St. Louis, reported from Oregon County in 1883 that a small group of birds had begun nesting. Forty netters were on-site and primed for action, but he feared that they would be stymied by the shooters, who were killing off the pigeons before they could nest. A local newspaper gave the total shipped to St. Louis from the site as more than ten thousand dozen, and that was two months before Judy's dispatch. This was the pigeon city that supplied the forty thousand birds that died before they could be shipped to hunting clubs.[2]

The last "really impressive" number of pigeons in Missouri appeared in January 1884. Crowds of pro and amateur pigeon catchers descended on the roost, located in the same area of Oregon County. Fortunately, one of the visitors published an account of his trip. With provisions enough for four days, his group set out for the pigeon gathering. After spending a restless night, they were up before dawn: "As the sun rose, the birds began to fly over us, all day at short intervals we were shooting right and left in the roost. The trees were literally crowded with them." As their outing concluded, they packed their gear and headed to the nearest town with a wagon full of pigeons. On the way they met more hunters, and as one small caravan they pulled into Augusta (now Thayer) with a combined total of 5,415 birds.[3]

Later in 1884, a nesting colony of three hundred pigeons took up residence in the wooded heights of Potter County, Pennsylvania, near Cherry Springs (now a state park but no longer a town). Remarkably, they completed their breeding without human interference, thanks to the secrecy maintained by the six or so persons who knew about it. Why they were so restrained in their speech and acts is not known, but that they allowed the pigeons to nest in peace may have been unprecedented in the passenger pigeon annals.[4]

The largest reported concentration of breeding birds in 1885 spread along the Oconto River in Langlade County, Wisconsin. But gunners drove the pigeons away before they could fledge any offspring. An hour after a flock of migrants showed up at Racine, five hundred hunters armed with shotguns appeared. The number of men may well have exceeded that of the birds. Longtime professional pigeon netter J. B. Oviatt found his last handful of nests that year in his home range of McKean County, Pennsylvania.[5]

The spring of 1886 saw remnant flocks of passenger pigeons converge on the upper Susquehanna basin of northern Pennsylvania to nest along the west branch of Pine Creek, where the beeches had produced a bountiful nut crop the preceding fall. John French had been watching the groups of birds as they passed over his hometown of Coudersport, and when he learned of their destination, he arose early one morning to see for himself a spectacle that was becoming less and less frequent.[6]

French began the thirty-mile trip before the sun had yet made its appearance, and not a single pigeon crossed his path over the entire hilly course. Nor were there any live pigeons when he finally reached the woods they had occupied. Instead, he was greeted by a sight so fantastic he might have been dreaming: "Young men were coming from the woods with bags full of dead birds. Many of them were lumberjacks, with high, spiked shoes on their feet; gray trousers, with legs chopped off at the knees, tucked into high-topped socks; mackinaw coats of bright red and brown and gray, in large checks; silken scarves around their necks; and high hats."[7]

The nattily attired dandies had decided to plunder the pigeons in style.

The timber company that owned most everything else in the area had just purchased the local general store and discovered an attic filled with silk hats dating from 1851, when the establishment first opened. They were of a design once marketed by P. T. Barnum and called Jenny Lind, in honor of the Swedish songstress with whom the shop's longtime owner had been infatuated.[8]

The night before French's arrival, the group of men with their fancy headgear had entered the pigeon city and begun firing. They did not stop until the woods were bereft of flying pigeons: "That was the death-blow to pigeons in Pennsylvania. They left in the night, which was clear, with a full moon, so the birds could see where to go in a northerly direction."[9]

What happened on Pine Creek had become the norm: as hunters invaded the nestings blasting away, birds would leave before fledging progeny. The connection between shooting and abandonment was perhaps first explicitly noted in 1869, when pigeons left a nest site in Wisconsin after being assaulted by armed attackers. Ornithologist Ludlow Griscom and others claimed that the major reason the species became extinct is that it was incapable of adapting to the ever-more-intense slaughter. But, in fact, the birds *had* changed: they were now less tolerant of these disruptions and would increasingly flee at their commencement.[10]

No one knows for sure where the retreating pigeons went when they left Pine Creek that night in 1886. It is possible that some relocated a short distance to the east and nested near Blossburg, on reforested timber land where "thousands of squabs were killed with poles in the little trees during the bark-peeling time." Others surmised that the birds, in their northward flight, traversed the skies of New York to settle somewhere in the remote vastness of Ontario or Quebec. One experienced pigeoner said that six hundred bred that year in a swamp near Lake City, Michigan. Naturalist Vernon Bailey found several nests along the Elk River, in Sherburne County, Minnesota, from the end of April to the middle of May. A tally of game killed in Missouri over the twelve months ending March 1, 1886, listed 8,129 "turtle doves" (presumably mourning doves) and 4,929 "wild pigeons." But the origins of the data are not given.[11]

The species tried to nest in at least two states in 1887. During late May and early June, groups of unknown size abandoned their efforts at sites near the Wisconsin towns of Wautoma and Sparta respectively. In both instances, the birds suffered harassment from shooting. The trappers had already arrived at Wautoma, but were once again deprived the opportunity to ply their trade by the thoughtless gunners. An unexpected location for breeding pigeons was Tyler County, Texas. Fourteen pairs nested and laid eggs, six of which were collected by Edmond Pope. One egg survives in the collection of New York's Columbia-Greene Community College, providing the only physical proof that the species did ever nest in Texas.[12]

By 1888, most ornithologists who thought about passenger pigeons probably lamented that the bird might well disappear without ever having been scientifically studied in the wild. When William Brewster, curator of birds at Harvard's Museum of Comparative Zoology, learned that big flocks of pigeons were settling in to nest near Cadillac, Michigan, he jumped at the opportunity to "learn as much as possible about the habits of the breeding birds, as well as to secure specimens of their skins and eggs." He grabbed a colleague and headed straightaway to Cadillac, where they arrived on May 8. To their great disappointment, however, they found that the flocks had moved on to points unknown. They were just a few weeks too late! But the netters, already poised for action, urged patience. The birds would certainly set down somewhere within reach; in due time that place would be discovered and the news would rapidly spread. After two weeks, however, the secret pigeon haunts remained hidden: "One by one the netters lost heart, until finally most of them agreed that the Pigeons had gone to the far North beyond the reach of mail and telegraphic communication." Indeed, those vernal flocks that had lured both scientists and pigeoners were the last large aggregations of passenger pigeons ever seen in Michigan. A mere decade had elapsed since Petoskey.[13]

COUNTDOWN TO EXTINCTION: 1890–99

Wild passenger pigeons still existed in 1890, but in pitifully low quantities. A. W. Schorger suggested a population of a few thousand. But as Brewster discovered, there was no place one could go and be assured of seeing the species in nature. The birds were too few and scattered. It seems, though, there was one place where one could see pigeons in the early nineties: the Boston market. During 1891, its purveyors received on one occasion twelve hundred birds from Pennsylvania and Missouri, and on another two thousand live individuals captured near Fayetteville, Arkansas.[14]

A number of reports come from 1892. While the pigeon supply dipped well below what could sustain national markets, early shoppers in Norfolk, Virginia, on January 16, 1892, could have indulged their taste for the species as both "wild pigeons and watermelons were among the delicacies exposed for sale in the city market this morning." A few birds supposedly nested in an old stand of hemlocks along Young Woman's Creek in Clinton County, Pennsylvania. The third record came from Bethel, Connecticut, where G. L. Hamlin watched seven birds in August and September as they regularly foraged in a tract of buckwheat. He shot one individual, but discarded it because it was in full molt and thus unlikely to make an attractive mount. Over the winter of 1892–93 "several hundred dozen" pigeons reached the St. Louis market from the Indian Territories. These may have been the last pigeons offered for sale in a big city, for a shipment arriving in New York City about the same time was condemned as being unfit for human consumption.[15]

Five reports of passenger pigeons were made during 1893. In late April, word came from Highgate, Virginia, that a few pigeons were breeding in the vicinity and, what is most remarkable, that an effort was under way to protect them. About the same time, twenty or so birds lingered in Maynes Grove, Franklin County, Iowa. One was discovered dead. A lone individual was taken at Morristown, New Jersey, on October 7. Also in the fall, a Mr. Riddick of Heywood County, Tennessee, saw eight birds, of which he shot one.[16]

A detailed and well-written account describes one of the two sightings from Illinois that year. To his great amazement Edward Clark discovered a male passenger pigeon in April 1893 as he birded in Chicago's expansive Lincoln Park. Clark was a writer on the staff of the *Chicago Tribune* and a graduate of West Point who rose to the rank of colonel. The pigeon rested on the branch of a maple, fully illuminated by the early-morning sun: "There were no trees between him and the lake to break from his breast the fullness of the glory of the rising sun . . . The sun made his every feather shine. Not a single feather was misplaced, and about the neck there was the brilliancy of gems."[17]

Clark had hoped to flush the bird so that it would fly to the north and out of the city. But, instead, to his dismay, the bird "winged his arrowy flight straight down the Lake Shore drive toward the heart of the city." But even if the pigeon had flown in the direction Clark thought safest, it might not have mattered; just a few months later in September, somebody shot a pigeon out of a flock of three near the Des Plaines River, in Lake County, a little to the north and west.[18]

Indiana hosted more pigeons during 1894 than it had in the previous two or three years. There were five sightings, including a flock of fifty in La Porte County. An even larger group, 150 in all, reportedly appeared in Whitewater, Wisconsin, one day in early May. Massachusetts lost its last known passenger pigeon on April 12 when Neil Casey shot a female at

Melrose. October 20 is the date on which a pigeon would intersect speeding lead in North Carolina for the final time. She met her fate in Buncombe County.[19]

With the dawning of 1895, the passenger pigeon as part of the North American landscape entered its final phase. Almost every record marked a "last." A question that needs to be addressed is by what standard should reports be included here. Schorger believed that the last records needed to be based on "specimens with credible data," a conclusion with which I agree. Sight records won't do, although with the endorsement of a local expert of known repute they are worth noting.[20]

The need for a corroborating specimen is particularly acute in the case of the passenger pigeon, which so closely resembled the widespread and common mourning dove. In numerous published instances, passenger pigeons miraculously became mourning doves once they were shot and examined in the hand. And sometimes even a bird in the hand or expert testimony was not enough to convince the true believer. W. DeWitt Miller and Ludlow Griscom of the American Museum of Natural History made an on-site investigation of purported passenger pigeons nesting in York County, Maine. The year was 1919, but "no reports that we have ever seen were so plausible or circumstantial, nor could we have encountered greater certainty in our correspondents." As further corroboration, one of the elderly witnesses was supposed to have been an experienced pigeoner in his younger days. But even after the two scientists observed the birds in question and identified them as mourning doves, the former hunter refused to concede. No doubt, to him they would always be passenger pigeons.[21]

During a cold spell in late January or early February 1895, two pigeons were shot at Mandeville, Louisiana, one of which is now in the collection of the Louisiana State University Museum of Natural Science. Pigeons continued to visit several places in Wisconsin that year. In Prairie du Chien hunters shot them. Michigan claimed ten birds sighted in West Branch and a young female taken on October 1 in a forested portion of Delta County. One flock of sixty and another of twenty-five spent some time in Indiana. Yet another pigeon ventured into Lake County, Illinois, this time reaching the tony suburb of Lake Forest. It was killed by a boy on August 7 using a rifle ball that mutilated the carcass. Fortunately, the young man brought the bird to a neighbor, John Ferry of the Field Museum, who, realizing its significance, preserved the skin anyway.[22]

Nebraska's final specimen fell to the Honorable Edgar Howard of Papillion, Sarpy County, when he shot it out of a flock of about twenty on November 9.

Another flock supposedly appeared later in the month at Omaha. Two birds became specimens in New York: a taxidermist recalled mounting a male killed in the spring at North Western, while another was shot in September at Clinton. A pair were noted in Grey County, Ontario, in mid-August, the last of the species the observer would ever see. Another bird shot that year, in Putnam County, West Virginia, is deemed the last reliable record for the state. When Oliver Jones collected a male, a nest, and an egg on June 21, 1895, near Minneapolis, he established the last verified record of both the species for the state and nesting in the wild anywhere. He would later donate the pigeon material and the rest of his extensive collection to the University of Minnesota's Museum of Natural History (the Bell Museum).[23]

A number of reports come from 1896, spread widely across pigeon range. Professor W. P. Shannon observed the wings of a bird taken at Greensburg, Indiana, either in the winter of 1895–96 or the following spring. No month or day is given, but market hunters are said to have taken three birds at Houston, Texas. Observers in Ontario reported two flocks, one on April 15 consisting of thirteen females or young birds, and one of eleven on October 22. Frank Chapman, editor of *Bird-Lore* and the creator of the Christmas Bird Counts, examined a mounted young female molting into adult plumage that had been obtained on June 23 in Englewood, New Jersey. Frank Rogers shot one near Dexter, Maine, on August 16. Wisconsin, once home of vast nestings, surrendered its second-to-last bird when a hunter shot a young male that rested on top of a dead tree. This occurred on September 8 near Delavan Lake, Walworth County, an area rich in acorns and grasshoppers, as evidenced by the contents of the pigeon's full crop. Iowa yielded its last passenger pigeon specimen when a male was shot near Keokuk on September 7 or 17. About a month later, on either October 23 or 25, the last fully corroborated specimen from Pennsylvania was taken north of Canadensis, Monroe County. Frank Cushing Norris shot the bird, an adult male, as it perched alone in a pine tree.[24]

The year 1896 would also mark the final recorded presence of the species in Louisiana. Just two days after Thanksgiving, A. E. McIlhenny, whose father invented Tabasco sauce, was pursuing prairie chickens in Cameron Parish. As he stalked his quarry, he came across a large group of mourning doves foraging in a field of corn and peas. Fraternizing with the doves was a lone passenger pigeon, which he shot. Hunting quail in Oregon County, Missouri, on December 17, Charles Holden encountered a flock of fifty pigeons skittering across the sky above him. He managed to collect two and sent them in the flesh to Ruthven Deane, the Chicago businessman who did

This photo of Albert Cooper and his three live passenger pigeon decoys from around 1870 appears in Henry Paxson's 1917 article on the species in Bucks County, Pennsylvania. It is one of only two known photos of live passenger pigeons that do not involve the flocks kept by Whitman or the Cincinnati Zoo. (From the collection of Garrie Landry)

Buttons, the second most famous stuffed passenger pigeon, is a young female collected March 24, 1900, at Sargents, Pike County, Ohio. Schorger proclaimed this to be the last passenger pigeon collected in the wild for which there was no doubt, but he was unaware of the males taken a year later in Illinois and two years later in Indiana. Buttons is on display at the Ohio History Center in Columbus. (Photograph by Steve Sullivan)

Museum dioramas provide the best opportunities to see passenger pigeons as they may have looked in nature. Created in 1955, this diorama from the Bell Museum of Natural History features a background painted by the noted nature artist Francis Lee Jaques. (Courtesy of the Bell Museum of Natural History, University of Minnesota, Minneapolis)

This example of the bird's image being used for commercial purposes was drawn by John Hintermeister (1869–1945) and copyrighted by Arm & Hammer baking soda in 1908. (From the collection of Garrie Landry)

This haunting painting, entitled *Beech Grove-Passenger Pigeon*, is by David Hagerbaumer, whose watercolors often feature game birds in flight. His works have been collected in two books, *Selected American Game Birds* and *The Bottoms*. (Courtesy of David Hagerbaumer)

Old-Fort Erie with the Migration of the Wild-Pidgeon in Spring is dated April 18, 1804, and was painted by Dr. Edward Walsh. A native of Waterford, Ireland, he served as lieutenant surgeon in the British Army and was stationed in various parts of Canada from 1803–1809. His pictures of Montreal, Fort Erie, and Fort George were released as aquatints in 1811. (From the collection of the Royal Ontario Museum, Toronto; painting data by Mary Allodi of ROM)

This tintype, probably taken in the 1860s but possibly as late as the 1880s, is one of only two known photos of live passenger pigeons that do not depict the flocks kept by Whitman or the Cincinnati Zoo. Two pigeoners pose with their gear and three stool pigeons. This picture has been slightly cropped on the top portion. (From the collection of Destry Hoffard)

Small Game of the Alleghenies is a stereopticon card that features an image unique in the passenger pigeon annals: along with gray squirrels and ruffed grouse, there are three freshly killed passenger pigeons. The photo was probably taken during the 1870s by R. A. Bonine. (From the collection of Destry Hoffard)

Passenger pigeons have been featured on the postage or commemorative stamps of such countries as Tanzania (shown left), Mozambique, Cuba (shown right), and Norway. (From the collection of Garrie Landry)

Male passenger pigeon drawn by K. Hayashi from a live bird in the collection of Charles Otis Whitman. Hayashi was among the few artists who painted the species using live passenger pigeons as his models. This and the other pictures in Whitman's book are considered among the most accurate drawings of the bird, although the Japanese-like background reflects the sensibility of the artist more than the reality of the bird's habitat. (From *Posthumous Works of Charles Otis Whitman*, plate 29 (1919), in the collection of Bowdoin College, courtesy of Susan Wegner)

The Travelers Insurance Company ran this picture as part of an advertisement in the September 1936 issue of *National Geographic*. (The original watercolor is part of the company's collection, but the artist is unknown.) The ad copy underneath the image was: "A century ago the great naturalist, James [sic] Audubon, observed this species in such abundance that he did not believe its numbers would ever be greatly diminished. Yet the last passenger pigeon vanished twenty years ago. If the fate of an entire species is unpredictable, how much more so are the fortunes of an individual? That is why the wise man, no matter how abundant his fortunes today, prepares for adversities that may befall tomorrow. Moral: Insure in The Travelers." (From the collection of Garrie Landry; scan by Steve Sullivan)

Shooting Wild Pigeons in Northern Louisiana is based on a sketch by Smith Bennett and appeared in the *Illustrated Sporting and Dramatic News* of July 3, 1875. Schorger considered this picture to be particularly accurate. (From the collection of Garrie Landry)

"Educational sliced puzzles" for children were produced principally by two firms, Layman and Curtsies and Selchow and Righter, for the purpose of helping children both spell and learn the names of animals. This particular puzzle of a passenger pigeon was part of a set that featured twelve other birds. (From the collection of Garrie Landry)

This painting by John Ruthven, often called the twentieth-century Audubon, depicts a male passenger pigeon at the John Roebling Suspension Bridge. Spanning the Ohio River from Cincinnati, Ohio, to Covington, Kentucky, it was the longest suspension bridge in the world when first opened on December 1, 1866. (Courtesy of John A. Ruthven at www.ruthven.com)

Mary Ijams holding a stuffed passenger pigeon in 1928 that her father, H. P. Ijams, had recently purchased. The bird had been shot by General Benjamin Cheatham near Nashville in 1856 and is now in the collection of the Great Smoky Mountains National Park. (Courtesy of Paul James, Ijams Nature Center, Knoxville, Tennessee)

Louis Agassiz Fuertes (1874–1927) was one of the country's premier bird artists of the early twentieth century. He produced a number of passenger pigeon paintings, including this one of an adult male, female, and young that appeared in Forbush's *Birds of Massachusetts and Other New England States* (1927). The plate also features two mourning doves, an adult and young. (From the collection of Garrie Landry)

John James Audubon's portrait of a male and female passenger pigeon is the best known of any for the species. His classic *The Birds of America* included 435 hand-colored plates and was released in four volumes between 1827 and 1838. These were accompanied by five volumes of text entitled *Ornithological Biography*. (From the collection of Garrie Landry)

Mark Catesby spent eleven years in the New World sketching and collecting natural history specimens before returning to England in 1726 and publishing *The Natural History of Carolina, Florida and the Bahama Islands*. Volume I contains the earliest known colored drawing of the passenger pigeon. The plate illustrated here is from a later German edition that relies upon Catesby's image. (From the collection of Garrie Landry and with additional information from Susan Wegner)

Alexander Wilson's *American Ornithology* (1808–1814) featured pictures of 268 species, including this passenger pigeon in volume V. The two warblers that accompany the pigeon are the Blackburnian (called by Wilson the "hemlock warbler") and the mysterious "blue mountain warbler," a bird of uncertain identity that has not been encountered since the early nineteenth century. (From the collection of the Morton Arboretum, Lisle, Illinois)

Passenger Pigeons in Flight was painted by Lewis Cross in 1937. Cross was a practitioner of "narrative natural history painting" and is perhaps unique in both drawing the species and knowing the bird in his lifetime. Narrative natural history painting is a term used by Walton Ford to distinguish art that is more expressive than art meant solely to depict a subject through a technically accurate portrait. (From the collection of the Lakeshore Museum Center, Muskegon, Michigan; photograph taken by Fred Reinecke)

This adult male passenger pigeon was shot on March 12, 1901, in Oakford, Illinois. It is the last wild bird known to be taken for which there is an extant specimen and for which no doubts as to provenance exist. It has been in the collection of Millikin University since 1947. A complete account of the specimen can be found in chapter 8. (Courtesy of Millikin University, Decatur, Illinois)

No. 4253 Chicago, Apl 25 1884

Acc't Sales by **Bond & Pearch,**

No 163 South Water Street

Of 2 Bbls

For Account and Risk of W. C. Waterman

Emery S. Bond.
Dan'l W. Pearch.

Received Apl 23

2 Bbls Pigeons 523 225 98 06

Ship two barrels – Will try to sell at 275

CHARGES:— Freight, Cartage, Commission, Net Proceeds to your credit.

Ex 2 70
4 90 7 60
90 46

Original sales slip from Chicago game dealers Bond and Pearch to professional passenger pigeon netter W. C. Waterman for the receipt of two barrels of pigeons (523 birds) captured near Madison, Wisconsin. The slip is dated April 25, 1884, and represents the last shipment of pigeons ever made by Waterman. (Courtesy of the Milwaukee Public Museum)

Photograph of Martha at the Smithsonian Institution's National Museum of Natural History, Washington, D.C. (From the collection of Garrie Landry; photographer unknown)

so much to document the last passenger pigeons. These birds are now among the holdings of the Chicago Academy of Science's Peggy Notebaert Nature Museum, and I can attest they are both stunning males. Chief Pokagon wrote to Deane in October 1896 that he had been "credibly informed that there was a small nesting of Pigeons last spring not far from the headwaters of Au Sable River in Michigan." Deane tried to seek corroboration by contacting the state game warden, Chase Osborn. Osborn did not know anything about the Au Sable nesting but mentioned other recent sightings and expressed the belief that a few birds did continue to nest in the northern reaches of the state.[25]

Although in 1897 a small handful of passenger pigeons were still eking out a living in the great outdoors, at most only two are known to have been killed that year. Harvard's Museum of Comparative Zoology has one bird that was taken on April 26 in Meridian, Wisconsin. The Chicago Academy of Sciences holds the other specimen, but the date of demise is less certain. The label lists the origin as Evanston or Rogers Park (that part of Chicago that borders Evanston) and the time as "around 1897."[26]

Margaret Mitchell accepted on good authority that in spring of 1898 a group of twelve to twenty pigeons nested in maple and beech near Kingston, Ontario. Their fate was not recorded, but at least eight pigeons did have fatal encounters with humans during that year. It seems a bit odd that of these eight birds, only one was taken in the spring. That occurrence was evidently never published, but fortunately the corpse wound up in the Harvard collection. Number 248528 died in Adrian, Michigan, on April 19, 1898.[27]

According to the *Osprey* ("An Illustrated Magazine of Birds and Nature"), sometime in 1897 or so an unnamed newspaper concocted a story that the Smithsonian Institution was offering generous remuneration for passenger pigeon specimens and/or information on where live birds could be found. Other papers disseminated the falsehood and it reached innumerable readers, many of whom responded with various claims and offers of their own. One cunning soul declared that he knew the secret hideout of numerous passenger pigeons but would divulge the information only upon first receiving payment. But amid all the bogus claims, there arrived at the Smithsonian an immature pigeon, killed on July 27, 1898, two miles east of Owensboro, Kentucky. In November another Kentucky passenger pigeon was reportedly shot in a hemp field near Winchester, but unfortunately the hunter cooked it and served it to his infirm wife for lunch. Afterward, he lamented his mistake and wished he had kept the feathers.[28]

In Michigan the lives of three passenger pigeons came to an end in

1898. The first two were a pair reportedly shot in July near Grand Rapids. No one would have known of this record had not the outdoor writer Emerson Hough mentioned it in his *Forest and Stream* column "Chicago and the West." Decades later, Schorger sought more information about the Grand Rapids birds and contacted the daughter of the man who had them in his possession at the time Hough wrote his article. Although she remembered the two mounts from her childhood, she could not tell him where they wound up.[29]

The final Michigan bird of 1898 was shot on September 14 and was for a number of years considered the last verified collection of a wild passenger pigeon. Nobody has questioned the identification or the date, but a major prevarication runs through the record. The event was announced to the world in two different notes published in 1903. (Note the five-year delay in publishing.) One of the papers was published in the *Auk* by James H. Fleming. He merely recorded the date, that the bird was immature, and the name of who mounted it. The second of the 1903 papers was penned by Philip Moody and appeared in the *Bulletin of the Michigan Ornithological Club*. He noted that Frank Clements shot an immature pigeon a few miles outside Detroit and that the bird was in Fleming's collection.[30]

Seven years later, J. Clair Wood published another article on the specimen, also in the *Auk*. He provided more details on the collection based on conversations he had had with Dr. Moody. Moody is quoted: "Mr. Clements and I were in the thick woods when we noticed three pigeons. They were flying above the tree tops, two abreast and the third behind and lower down. The latter bird lit near the top of a tall tree but the others continued their flight without a pause. I could have shot it but thought it was a Mourning Dove." A careful reading of the article reveals that Moody never says explicitly that Clements shot the bird, but since Moody did not and someone did, that can only be Clements. All was well and good until 1951 when Norman Wood (Clair's brother) published *The Birds of Michigan* and informed his readers that Frank Clements is a pseudonym for Philip Moody. What is going on here? Did Moody lie in his first paper and then later to Clair Wood? At least part of the answer can be found in the statute Michigan enacted in 1897. Under this law, Michigan became the only state to ban all killing of the passenger pigeon. Thus killing a pigeon in 1898 within the borders of Michigan would have been in violation of state law. That is clearly why the announcement of the specimen wasn't made until 1903. And even at that date Moody was obviously not comfortable admitting that he broke the law, so he created Frank Clements. But Moody's motivation for getting the story republished in the *Auk* twelve years after the event and repeating the falsehood remains obscure.[31]

On September 14, a male passenger pigeon was taken by Addison Wilbur at Canandaigua, New York, in the Finger Lakes region. Elon Eaton was present at the shooting and describes the individual in his *Birds of New York* as one fledged the preceding spring, as it was not yet in full adult plumage. Canada's last confirmed pigeon was a male whose life ended in Winnepegosis, Manitoba. According to George Atkinson, who stuffed it, the date of its demise was April 10. The bird was "in the pink of condition in every way" and was the "only specimen [he] was ever privileged to handle in the flesh in Manitoba."[32]

The final specimen from 1898 has not hitherto been part of the passenger pigeon literature. The source of the record is Amos Butler, the grandfather of Indiana ornithology, whose monograph *The Birds of Indiana* (1898) served as the definitive reference for decades. He submitted most of his articles on Indiana birds to the *Proceedings of the Indiana Academy of Science*, including notes published in 1899, 1902, and 1912. Sandwiched between them, "Some Notes on Indiana Birds" appeared in the *Auk* in 1906. It seems likely that Schorger and other non-Indiana ornithologists assumed that the *Auk* paper and the 1897 book revealed all that Butler had to say about passenger pigeons in his state. But by overlooking the three papers in the *Proceedings* they missed two significant records.

Butler includes reports from a variety of people but makes clear those specimens that he himself saw and identified. In the 1899 paper he provides the details of one recently collected bird: "One mounted by Beasley and Parr of Lebanon, was killed by Frank Young, Wilson P.O., Shelby County, Indiana near that place around September 24, 1898. It was in company with two doves in a patch of wild hemp when found. The specimen is in the possession of W. I. Patterson, Shelbyville, Indiana." He mentions the record again in the 1902 paper, reiterating the September 24 date and specifically stating, "I have seen this specimen."[33]

Three reports emerge from 1899. The most poorly documented of them is the mere mention in *Birds of Western Pennsylvania* that Dr. McGrannon shot one at Roulette, Potter County, sometime during that year. Next, in order of increasing documentation, concerns the Little Rock, Arkansas, merchant who received a male pigeon in with a load of quail that arrived from Cabot, Arkansas. He knew the significance of the bird, for he displayed the specimen for a few days with a sign saying it was the last of the species. The time was late December. The report boasting the best credentials of the three is of a young pigeon that was shot in Babcock, Wisconsin, between September

9 and 15. Schorger considered it the last reliable record from the state and published a note in the *Auk* elaborating upon the initial report. Emerson Hough was part of the group that secured the specimen and averred that he still had the skin in his possession at least as late as 1910. Where it eventually wound up is unknown.[34]

STILLED WINGS: THE NEW MILLENNIUM AND THE END OF THE PASSENGER PIGEON AS A CREATURE OF THE SKY

"Who killed Cock Robin?"
"I," said the sparrow,
"With my bow and arrow,
I killed Cock Robin."
"Who saw him die?"
"I," said the fly,
"With my little eye,
I saw him die."

—ANONYMOUS, NINETEENTH-CENTURY ENGLISH RHYME

A first-of-the-year female, Buttons, or the Sargents' Pigeon as preferred by some purists, is one of the best-known wild birds that North America has produced. Her popular sobriquet refers to the objects that were used by the taxidermist to fill the holes on the sides of her head when the preferred glass eyes were not available. She figured as the main character in Allan Eckert's 1965 novel, *The Silent Sky: The Incredible Extinction of the Passenger Pigeon*, where she was called an adult "he" who had a white splotch on one wing. Christopher Cokinos has written at length about his search for the young man who allegedly shot her, and the dedicated scholar Geoffrey Sea, whose house is next door to where the bird was killed, has spent years ferreting out the details about those involved. What brought these people and others to focus on this particular specimen was Schorger's proclamation "I am willing to accept as the very last record the specimen taken at Sargents, Pike County, Ohio, on March 24, 1900."[35]

The saga of Buttons began with a 1902 article by W. F. Henninger on the birds of "Middle Southern Ohio." His passenger pigeon account takes four sentences, the last two of which are relevant here: "On March 24, 1900, a solitary individual was shot by a small boy near Sargents, close to the boundary line of Pike and Scioto Counties, and mounted by the late wife of ex-sheriff C. Barnes of Pike County. This is the only authentic record for twenty years." A Lutheran minister by profession, Henninger resided in nearby Waverly and was widely respected as an ornithologist and a taxidermist. He failed to comment on the gender or age of the bird, which has led Sea to conclude he never examined it. If that is true, it makes what follows even more critical.[36]

If, indeed, the identification rested with Barnes, the record would have warranted no more than a quick mention in any history of the species. But fortunately, Clay Barnes offered to donate the bird to the Ohio State Archaeological and Historical Society in October 1914, the month after the last captive bird died. The original accession card is now missing and the documents that remain are ambiguous on the details of the acquisition. But that card was quoted in a 1955 story that appeared in the *Ohio Conservation Bulletin.* It was dated February 25, 1915: "Received from Clay Barnes Waverly. Killed by Mr. Barnes himself." The article relied heavily on the views of Dr. Edward Thomas, who served as both curator of natural sciences at the Ohio State University Museum from 1931 to 1962 and the curator of natural history at the Ohio Historical Society. No one knew the collection better than Thomas. Based on the accession card and other factors less clear, he evidently believed that Barnes had indeed shot the bird and invented the small boy to avoid any disapprobation (shades of Dr. Moody/Mr. Clements). The *Waverly Courier Watchman*, spurred by the recently passed Ohio law protecting game, had been railing against those who use "rifles in shooting every bird they see," and thus it was politically inexpedient for the good sheriff to admit that he fell into that category.[37]

On the other hand, Thomas assumes that Henninger spoke with the Barneses about the incident. He would have had but a narrow window in which to conduct that conversation, as the bird was shot in late March and Mrs. Barnes died around July 8, just three days before her newly born infant also passed away. (Sea believes mother and child died of arsenic poisoning, as the substance was commonly used in the preparation of specimens.) But if that interview had occurred and the Barneses had made up the boy story, they would have been lying to Henninger's face or Henninger agreed to go along with the ruse. Sea, though, does not think this talk did ever take place, because of both Mrs. Barnes's death and local politics: divisions within Pike County at the time would likely have separated the Barneses from the Henningers.

Decades later, Christopher Cokinos tracked down the supposed shooter, Press Clay Southworth. Southworth had died in 1979, having lived just six years shy of a century. Cokinos describes him in the most glowing terms, calling him "a strong and polite man, witty and usually patient," who grieved "deeply" for two wives who predeceased him. When Southworth read an article in *Modern Maturity* magazine in 1968 recounting the Buttons tale, he responded with a letter to the editor giving an extremely detailed recollection of events that took place almost seventy years earlier. The youngster had been tending the cattle in the barnyard when he saw "a strange bird feeding on loose grains of corn." The bird flushed and landed in a tree. The visitor puzzled Southworth, for it was too big for a mourning dove and flew differently from a rock pigeon. He ran into the house to get permission from his mother to dispatch the bird with the family's 12-gauge shotgun: "I found the bird perched high in the tree and brought it down without much damage to its appearance. When I took it to the house, Mother exclaimed—'It's a passenger pigeon.' " His father concurred in the identification and urged him to take the bird to Mrs. Barnes, who prepared the mount, improvising when she discovered she had no glass eyes left.[38]

As far as I know, no one since Schorger has examined the question of which record represents the last wild bird. It is, of course, impossible to know when the last wild passenger pigeon died. Buttons, after all, was a young of the year, so it is perfectly possible that one or both of her parents were still around, and if they could have fledged young, so likely could others. Most of these reports, including Buttons, had been first challenged by James Fleming, who seemingly tried to cast doubt on the timing of all specimens collected after September 1898, the date of the most recently killed pigeon in his possession.[39]

In my research, I have come upon two records unknown to Schorger that extend the confirmed presence of the species in the wild to 1902. One of these is a specimen at Millikin University in Decatur, Illinois, which is the last undoubted passenger pigeon known to have been collected in the wild for which there is an extant specimen. From Paul Hahn's *Where Is That Vanished Bird?*—an amazing work that gives the location of every known passenger pigeon specimen in the world (along with a few other extinct or nearly extinct birds)—I knew that a passenger pigeon was in the collection at Millikin. A mutual friend contacted David Horn, the ornithologist at Millikin, about my interest, and Horn sent me several photos of the bird in its display case. One shot clearly revealed the label: "This bird was killed in March 25, 1901 near Oakford, Illinois and mounted by O.S. Biggs, San Jose, Illinois." On the back of the case were the same words with some additional information. Biggs had mounted the bird for M. O. Atterberry, but then purchased it from Atterberry a short time later. The last sentence states, "This was the last Passenger Pigeon killed in Illinois that there is any record of." And it ends with "O.S. Biggs." I was flabbergasted to see a specimen that was purportedly taken so late.

Further research revealed additional details. The front page of the August 1901 issue of the *Oologist* (vol. 18, no. 8) includes this short note: "Under date of March 25, Mr. O.S. Biggs of San Jose, Ill., writes: 'A friend sent me a fine specimen of a male Passenger Pigeon which was killed March 12 near Oakford, Illinois. It is the first one I know of being killed here in 8 or 9 years. I have it mounted and in my collection." For reasons that are not clear, another central Illinois naturalist and friend of Biggs's, R. M. Barnes, sent a similar letter to the *Oologist* that was published in April 1923: "We are informed by O.S. Biggs of San Jose, Illinois, that one of these birds [Passenger Pigeon] was killed March 25, 1901 at Oakford, Illinois, and was mounted by him and is still in existence. Owing to the lateness of the date, we thought this capture worthy of record." Barnes obviously did not know that Biggs had sent the same information to the same journal twenty-two years earlier. The only real contradiction is that the original note as published gave the killing date as March 12 and the letter date as March 25, while Barnes and the label on the back of the case say the bird was shot March 25. I would be inclined to accept March 12, 1901, as the date the bird was shot and March 25 as the date Biggs sent his letter to the *Oologist*, particularly since it is not clear who wrote the label. But at most the discrepancy is thirteen days and is based on a plausible reason, confusion between the date of Biggs's letter to the journal and the date of the actual shooting.

Nothing is known about the circumstances of the bird's killing or whether Atterberry was even the actual shooter. But both Atterberry and, especially, O. S. Biggs attained enough prominence to have left a historical record. Atterberry and a partner bought what had been a drugstore in Oakford, Menard County, in 1892. Later, Atterberry acquired sole ownership in what would become a highly successful business that offered a wide variety of services and products including banking, hardware, drugs, groceries, and sundries. Residents referred to it as Sears Roebuck of Oakford. An operation such as that would have attracted local hunters and others who had unusual items to show or sell, so if Atterberry was not the shooter himself, he would have been in a good position to have acquired a freshly killed passenger pigeon.[40]

Biggs (1861–1947) was a college graduate who came from one of the founding families of Mason County. He had eclectic interests and was at various times occupied as a pig farmer, banker, tax assessor, apiarist (vice president of Illinois River Valley Beekeepers Association), and taxidermist. This last pursuit is the most relevant to this story. He donated specimens to the Illinois State Museum and "reportedly" had some of his work displayed in such universities as Harvard, Yale, and Chicago. A note published in the *Auk* on one of Illinois's few prairie falcon records credits Biggs as the source of the specimen.[41]

Biggs had two daughters, Hazel and Olive. When he died, his extensive natural-history collection was divided between them. Hazel, who also became an accomplished taxidermist, used her home as a small museum that featured the mounts created by her and her father. These holdings were eventually donated to the Illinois State Museum. Olive, a liberal arts major who graduated from Millikin in 1926, wound up giving her birds to Millikin. In the college *Bulletin* of July 1947, there is a discussion of the contributions collected to help finance a proposed Science Hall: "More than 200 birds and animals were given by Mrs. Cyril Gumbinger (Olive Biggs) '26. It is a collection her father made through a lifetime, including white owls, eagles, and a passenger pigeon now extinct. It is valued, we are told, at 'several thousand dollars.'"

The second record I wish to introduce refers to an Indiana specimen for which all salient facts are known and the identification is beyond dispute; unfortunately, however, the bird itself was destroyed many decades ago. Details appear in those obscure notes of 1902 and 1912 that Amos Butler published in the *Proceedings of the Indiana Academy of Science.*

The bird was shot on April 3, 1902, and Butler learned about it shortly thereafter "through the kindness of Fletcher Noe," who operated a pawn/taxidermy shop in Indianapolis. Butler quotes a letter from Charles K. Muchmore, a pharmacist in Laurel:

> The bird, which is a beautiful male, was taken by a young man named Crowell, near his home, about two and one-half miles southwest of this place. He reported that there were two. He heard the bird cooing and shot it and brought it to me, having concluded that it was something new. You can imagine how we almost took it away from him when he unrolled it out of a bloody old newspaper and began to inquire if we knew what it was . . . I

> used to see them occasionally in Iowa about 1882–3, and although I was then very small, the specimen was not new to me, and I, of course, at once recognized the same.[42]

If matters ended with that letter, the record would rest on Muchmore's identification. Butler quotes him in a number of his papers so he likely was reliable, but his stature would not be enough to carry the weight necessary to establish the veracity of the report. Unfortunately, he never donated the specimen to a collection where it can currently be examined, but he did the next best thing. Butler continues the story in his 1912 paper: "The last verified record for this State is from Franklin County. Two birds were seen, and one was shot, near Laurel, April 3, 1902. The specimen taken was submitted to the writer for verification and was returned to Mr. C.K. Muchmore, the owner, at Laurel." It is inconceivable that Butler would have misidentified a dead adult-male passenger pigeon that he had the opportunity to study in hand at leisure.[43]

I wanted to find out more, but inquiries in Laurel yielded nothing. Might there be a hint of the bird's fate buried in Butler's writings? Don Gorney, a longtime birder who enjoys a wild-pigeon chase, became my accomplice on a trip to the Lilly Library at Indiana University in Bloomington, where most of Butler's papers are housed. We spent a day perusing the contents of many boxes and realized that almost all of it related to Butler's social-welfare and prison work. There was, for example, the case of the two young women who tried to bribe their way out of incarceration at a state facility by offering a gardener the only thing they could provide that he felt was of value. They kept their part of the bargain but he reneged on his. An old story, I suppose. But we found virtually nothing about birds.

Undaunted, Don started poking around and learned that the papers we sought were in the possession of the Indiana Audubon Society and kept at their headquarters. The society then passed the material along to their archivist and board member, Alan Bruner. Don contacted him, and Alan graciously invited Don and me to his house, where we could go through two huge boxes of unorganized Butler stuff. Alan lives near Turkey Run State Park, which is three hours from me and two hours from Don. The day I selected turned out to coincide with the season's first serious discharge of snow. To arrive in advance of the precipitation, I left the day before and stayed at the closest motel. In the morning, the roads slick with slush, I slowly made my way around the S-curve, over one bridge, past the cemetery, and then across the final bridge before reaching Alan's house. He and his wife, Jackie, answered my knock and the day began. Don arrived shortly thereafter.

Amos Butler. Courtesy of Prints and Photographs Division, Library of Congress

Alan had moved the two big boxes next to a table off the kitchen. The boxes brimmed with papers of various kinds in no order. Both Alan and Don explained how many hands the contents had gone through before reaching their current state. I was not hopeful.

Amazingly, however, Alan found them: two letters from Muchmore to Butler, one dated August 30, 1932, and the other just a few weeks later on September 19. We learned that Muchmore obtained the pigeon from Crowell in exchange for "a week's supply of tobacco trade" (September 19) and that Muchmore stuffed the bird himself. Crowell had since moved to Nebraska.

Then there was the August 30 letter, which brought closure to the affair, albeit in a disappointing way:

> I am very sorry to report that the pigeon was destroyed some 17 years ago, in this wise: I was taken with tubercular trouble and dropped everything and headed for the mountains. I was in Grant County at the time. Later, I wrote to a friend to go to the store and get my specimens and take them home with him until I might be able to call for them. This he did, but unfortunately his wife promptly threw them into a woodshed attic and the winter rains beating through the roof wrecked them all, so that months afterward when I inquired about them they were gone. I shall always regret my failure to put this specimen in the state museum as you suggested and as I had fully intended to do.

(There is probably a lesson in this for all of us.)

This record is unassailable and is the strongest claimant as the last wild bird killed. Thus ends, to the best that can be ascertained, the existence of the passenger pigeon as a wild bird. But a hundred years ago, a few people were not yet ready to concede the fact. Schorger describes what follows this way: "No better example of eternal hope, so characteristic of man, can be found than the search for a living wild passenger pigeon long after it had ceased to exist."[44]

THE FINAL SEARCH

Sight records and even some claims of dead passenger pigeons continued for years. None other than Theodore Roosevelt felt certain that he had observed a flock of seven on May 18, 1907, at his presidential retreat, Pine Knot in

Albemarle County, Virginia. Roosevelt obviously knew the status of the species and quickly penned letters to two confidants. The most detailed exposition of his experience went to C. Hart Merriam: "I saw them flying to and fro a couple of times and then they all lit in a tall dead pine by an old field. There were doves in the field for me to compare them with, and I do not see how I could have been mistaken." The other letter went to his close friend John Burroughs, the famed nature essayist. To Roosevelt's delight, Burroughs responded that pigeons had also been seen in New York and on occasion in sizable numbers.[45]

Burroughs had independently concluded that "a large flock of wild pigeons still at times frequents this part of [New York], and perhaps breeds somewhere in the wilds of Sullivan and Ulster County." His view drew the attention of William Mershon, who was preparing his book at the time. Mershon wrote him and asked if the observers had possibly mistaken plover, teal, or other migratory birds for pigeons as he himself had done. Burroughs replied that his view was based on the accounts of several people he knew well, one of whom said the birds spread across the sky at a length of a half mile. Still, Burroughs was enough of a scientist to harbor at least a trace of doubt for he commented that with all the pigeons in upstate New York "they ought to be heard from elsewhere—from the south or southwest in winter." He further acknowledged that their presence raised another question as well: "If these flocks were pigeons, where have they been hiding all these years?" Where, indeed?[46]

Another person who still held out hope was the eminent scientist Clifton Fremont Hodge of Clark University in Worcester, Massachusetts. Called "a born naturalist," Hodge possessed strong interests in a variety of fields including forestry, ornithology, physiology, and environmental education, long before it was called that. Hodge was tending his garden one fall morning in 1905 when a flock of about thirty long-tailed birds flew over. They were so close, he harbored no doubt as to their identity. "The passenger pigeons are not extinct," he yelled as he waved his hat excitedly. Like others before him, he read meaning into the presence of the bird, although the message he divined was unique: by granting him this privilege, the fates wanted him to prove to the ornithological community that the species still survived outside a cage.[47]

During the American Ornithologists' Union meeting in the fall of 1909, Hodge presented a paper he called "The Present Status of the Passenger Pigeon Problem." It was a call for action. He asked the assembled bird scientists, "Do you think that a scientifically adequate search has been made for

Ectopistes migratorius?" Nobody answered strongly that such an effort had been mounted, and some, including the influential C. Hart Merriam, replied firmly that it had not. Hodge went on to decry the offers made by various parties that they would pay for any recently killed pigeons. Putting a bounty on whatever few remaining birds there might be would surely minimize their chances of survival. Rather, he thought, the rewards should be for active nests that could be verified by competent ornithologists.[48]

After the session, Hodge had a conversation with Colonel Anthony Kuser, one of those who had offered money for a dead bird, in his case $100. Hodge was conciliatory, saying he did not expect everyone to concur with him, but to his surprise, Kuser agreed completely with his position. When Hodge then suggested the reward should go for information about live birds, Kuser not only obliged but also raised the ante to $300. Ultimately, the joint announcement stipulated these conditions: "Three hundred dollars for first information of a nesting pair of wild Passenger Pigeons undisturbed. Before this award will be paid, such information, exclusive and confidential, must be furnished as will enable a committee of expert ornithologists to visit the nest and confirm the finding. If the nest and parent birds are found undisturbed, the award will be promptly paid.[49]

Other affluent fans of the passenger pigeon soon followed Kuser's lead: William Mershon for the first nest or nesting colony in Michigan, $100; Professor C. O. Whitman and Ruthven Deane for first nest in Illinois, $100 ($50 each); John Childs for a nest anywhere in North America, $700; A. B. Kinney for first nest in Massachusetts, $100; Edward Avis for first nest in Connecticut (will confirm at his own expense), $100; John Thayer for first nest in any state not yet spoken for up to five, $100 per state; and Allen Miller for first nest in Worcester County, Massachusetts, $25. The contributors and amounts varied a bit over time. These offers and specially prepared colored plates of passenger pigeons were distributed widely in the hunting journals and local newspapers.[50]

People realized that locating living birds would be but the first step. Hodge proposed that if the search proved successful, there should be established a Passenger Pigeon Restoration Club, which would reach out to public and private agencies to provide protective laws and "warden service." The goal would be to ensure security for the birds over their entire range: "The organization of the people of a continent around such an interest is in itself an inspiring thing."[51]

Reports poured in. Nests also found their way to Hodge, happily mostly of mourning doves, but at least one of either a sharp-shinned hawk or merlin.

He lamented that the press was not more careful in emphasizing that payment would only be made for active passenger pigeon nests. George Harrington of Waltham, Massachusetts, did not even wait for official confirmation before going directly to the newspapers with his discovery of a pair of birds in an unnamed town. They were cooing away in a large oak. While everything that Hodge published referred to a maximum award of around $1,000, Harrington was expecting closer to $10,000. It seems he had the idea that both the federal and state governments had chipped in thousands of dollars as well. At the time of the article, Harrington was waiting to hear back from Hodge, so he could show him the pigeons and collect his small fortune. Hodge's failure to even mention the incident suggests that the passenger pigeon claim was no more real than the expected cash windfall.[52]

But enough sight records, albeit every one lacking details, came in to warrant continuing the search one more year. During 1911, four reports came in that led to field investigations, but no surprises. A number of eyewitness accounts did seem highly credible. In two of these, birds were shot, but in each instance the evidence wound up in a pot. Even the feathers vanished. Hodge called this "the nightmare of the whole situation": whether they were correct in their identifications or not, people continued to kill birds they thought were passenger pigeons.[53]

Then in October Hodge received a letter from one Dwight Cushman of Hebron, Maine: "One day recently, while out hunting, I shot a bird and had it mounted by one of our leading taxidermists. It proved to be a Passenger Pigeon (*Ectopistes migratorius*). I think it is a young bird as it has dark spots on the back. Please reply giving me some more information concerning this bird." Hodge, probably quivering with anticipation, promptly mailed off a package that included "leaflets with photographs and underscored boldly in red ink the comparative lengths of the Pigeon and Mourning Dove. I also enclosed . . . colored plates of the two birds." In his cover letter, Hodge told Cushman to compare the enclosed with his bird. If he still thought the original identification was correct, Cushman should send the bird express and he would be compensated for the expense. "An early express brought a little box," inside of which was indeed a bird. Hodge sent the mourning dove right back to Maine, charging Cushman the eighty-seven cents for the return express. This incident illustrates so well why an acceptable late record of a wild passenger pigeon had to be either of a specimen examined by people competent to identify it or live birds seen well by multiple ornithologists.[54]

Hodge did not give up easily and the offer was extended one more year. By now the numbers of contributors had increased. Henry Shoemaker said he would give $200 to the first nest in Pennsylvania and would throw in $25 more if it was protected. A nest in New York would bring $100 courtesy of John Burroughs. But Hodge made it clear that this was it: if no confirmed passenger pigeon nestings materialized by October 31, 1912, all the offers would be rescinded. Halloween came and went. There were no treats. The scientific community had given it its best shot, but they were too late by years.[55]

All the passenger pigeons in the world that still drew breath resided in a cage in the Cincinnati Zoo. And her name was Martha.

CHAPTER 9

Martha and Her Kin: The Captive Flocks

By 1900, there were three captive flocks of breeding passenger pigeons. Martha was the progeny of one, born in a second, and spent most of her life in the third. The entities that maintained them, an interested amateur, a distinguished academic, and the nation's second-oldest zoo, were very different, as were the reasons they acquired and kept the birds in the first place. If events had proceeded differently, Martha might not have been the last of her species. But what-if runs through this entire saga and can never be answered.

Before considering the three well-documented captive flocks, I wish to linger on the little-known account by Henry Shoemaker that hints of other possible captive passenger pigeons. The story goes that one Jack Kreamer of Montoursville, Pennsylvania, kept ten stool pigeons in his cabin on Loyalsock Creek. By New Year's Eve 1908, only three remained, and a cat did away with one of them sometime during that long winter's night: "The old man, despairing of the return of the 'vanished millions,' hastily killed the two survivors and had them mounted."[1]

Shoemaker, writing this around 1919, goes on to reference persistent rumors that similarly kept pigeons might still survive in a few isolated homesteads secreted in the mountains of Pennsylvania. He had the resources and influence to track them down, so the ensuing silence resolves the matter. But I find it a haunting notion nonetheless that the last passenger pigeon might not have died in a modern zoo but a ramshackle barn where a faltering old man tried to preserve a living connection to his youth.

MILWAUKEE

One of the captive passenger pigeon flocks was created and maintained through the efforts of David Whittaker. As a young man, he and his wife, Maria, searched for gold in California and then the Yukon. After more years prospecting in Ontario and the upper Midwest, they eventually settled in Milwaukee, on a bluff above the Milwaukee River, where they established a swimming pool and a school. Many of the city's leading citizens, both men and women, became students. His most accomplished swimmer, though, proved to be his son George, who set world-record times for long and middle distances at the 1893 Columbian Exposition in Chicago.[2]

Wandering around the wilderness looking for minerals enabled him to develop and strengthen his fascination with nature. He evidently spent a lot of time looking up as well as toward the ground. Once he settled down, he sated this long-standing avian interest by raising birds. In what I have read about him, only two species are mentioned. He kept a flock of about twenty "bluebill wild ducks" (probably greater or lesser scaups). They would spend the day in the riverine marshland by his home doing what fertile ducks do, then return in the evening for food. In most years, ten to twelve young would be added to the group. What he did with the extra ducks or when the river froze is not recorded.[3]

Far more has been written about Whittaker's small aviary that housed his passenger pigeons. Wire netting composed two sides of it, while the top and the two remaining sides were of glass. Branches and poles provided perches, and a couple of shelves supported the nests that would eventually be built. The first occupants arrived in the fall of 1888, when Whittaker obtained two pairs of passenger pigeons captured in Shawano County, Wisconsin, by a "young Indian." One pair was of mature birds, while the other was of birds less than a year old. Unfortunately, the two older pigeons did not last long: "One . . . scalped itself by flying against the wire netting and died," while the other escaped.[4]

Whittaker proved himself a patient and skilled aviculturist. After several fruitless years, his flock of two began to grow, eventually reaching fifteen. The slow, tedious process required close observation and care. Impediments to success included the destruction of nests and eggs by the female and other members of the flock, as well as the killing of squabs by older males. During his viewing of Whittaker's facility in March 1896, Ruthven Deane witnessed the early stages of nesting:

> When the pigeons show signs of nesting, small twigs are thrown on to the bottom of the enclosure . . . There were three pairs actively engaged. The females remained on the shelf, and at a given signal which they only uttered for this purpose, the males would select a twig or straw, and in one instance a feather and fly up to the nest, drop it and return to the ground, while the females placed the building material in position and then called for more."[5]

When the adults fed young, Whittaker showed great astuteness by augmenting the regular diet of grain with worms and insects. These high-protein foods were released into a box of soil: "At times the earth in the enclosure is moistened with water and a handful of worms thrown in, which soon find their way under the surface. The Pigeons are so fond of these tid-bits, they will often pick and scratch holes in their search, large enough to almost hide themselves."[6]

Even after lives of confinement, the birds remained timid and wary. To observe them closely, Deane found, one had to approach the cage slowly and deliberately, lest they scatter in a panic. When storms approached, traits born of their genetic inheritance manifested themselves in yet another way: "The old birds will arrange themselves side by side on the perch, draw the head and neck down into the feathers and sit motionless for a time, then gradually re-sume an upright position, spread the tail, stretch each wing in turn, and then, as at a given signal, they spring from the perch and bring up against the wire netting with their feet as though anxious to fly before the disturbing elements."[7]

CHICAGO

Now the scene shifts a hundred miles south, to the University of Chicago. When John D. Rockefeller helped underwrite the founding of the school in 1890, he in essence gave the first president, William Rainy Harper, a blank checkbook with which to build the institution. Harper stocked his new uni-versity in the same way George Steinbrenner replenished the New York Yankees year after year: he bought key players from other teams. One of Harper's successes was enticing biologist Charles Otis Whitman to leave Clark Uni-versity and come to Chicago. Whitman was one of those rare academicians who published relatively little but became highly coveted through his activi-ties and his students.

Charles Otis Whitman at his Chicago aviary in 1900 with his various pigeons and a northern flicker. Wikimedia Commons

Born in Maine in 1842, Whitman seemed to lack interest in most everything as a young child, for one researcher says of him, "There is no hint from any of those I interviewed that young Whitman had any desire to play, draw, paint, climb mountains, travel, or build boat, engine, or carriage." He was rescued at the age of twelve from a life of apathy by developing a strong interest in birds, a sure sign of exceptionalism. When his pet blue jay died, he stuffed it. Additional collecting of specimens ornithological and mineralogical transformed the family home into a museum of sorts. A cousin recalled that Whitman once endured pelting rain for hours as he stood in a pond with his gun awaiting the return of a bird he wanted for his collection. While he was at Bowdoin College, in Brunswick, Maine, all anyone recalled of him was that he devoted his spare time exclusively to the collecting of birds.[8]

Whitman received a doctorate from the University of Leipzig and taught at the Imperial University of Tokyo. He then spent some time at the Zoological Station in Naples, Italy, before returning to the United States, where he directed a laboratory in Milwaukee. Whitman resumed his teaching career at Harvard, Clark, and Chicago. His research delved into such fields as animal behavior, embryology, evolution, and anatomy. But though he was an energetic scientist, he withheld publishing much of his work because he felt it would be premature unless he nailed down every detail and examined every reference. In the meantime, Whitman founded and edited the *Journal of Morphology*. He also helped establish and was the first director of the Marine Biological Laboratory at Woods Hole in Massachusetts, an effort to create a community of scientists who could study living organisms in an unfettered setting. Or, as one of its major funders described it, Woods Hole would have "many elements of a biology club."[9]

One of Whitman's true loves was pigeons. As a youth, instead of playing or building carriages, he used to watch "them by the hour." As a mature scientist, he amassed a large collection that eventually reached 550 live birds of around thirty species. These were housed in small cotes situated around his house in Hyde Park, a few blocks from the university. At least a few were always kept inside his home, serenading the human occupants with constant cooing. His goal was to study behavior, evolution, and genetics, and by working with such a wide variety of subjects he hoped to buttress his support of orthogenesis, a view that holds evolution to be "a directed and progressive process" determined by the characteristics of the species rather than the more random effects imposed by environment. This position had few adherents while he was alive, and during the years it took for his works to be published posthumously, its vitality had ebbed to the point where his findings were largely ignored.[10]

On March 4, 1896, Whitman added one more species to his collection when he bought three passenger pigeons from David Whittaker. By the following year, he would have in his possession all fifteen of the Milwaukee birds. Although nine chicks hatched, only four survived. With nineteen birds, Whitman returned seven to Whittaker. At least that is the most frequently related account. A very different story of how Whitman obtained his pigeons was told by Whittaker in a 1928 article. In this version, Whittaker's entire flock of 18 was stolen while Whittaker was out of town and presented to Whitman who purchased the lot for $1,500, apparently without knowing that he was buying hot property.[11]

Over the next few years, the overall number of pigeons in Chicago increased so that by the beginning of 1902 there were sixteen, evenly divided between males and females. In addition, there were two hybrid birds, offspring of a male passenger pigeon and female ringed dove. Both males, they proved to be sterile. Whitman noted the hardiness of the passenger pigeons: they fared well in their outdoor enclosures and resisted disease much more effectively than many of their neighbors. He called the species "my special pets."[12]

One of Whitman's experiments sought to determine how much of a bird's behavior was dictated by unvarying instinct versus its capacity to adjust to changed circumstances. He compared three species and found each represented a different "grade." When he approached a passenger pigeon nest to take the egg, the hen moved away but soon returned. She looked into the nest and reentered as if the egg were still present. But after a short while, she sensed that something was missing. She then abandoned the nest without making any attempt to recover the egg placed close by, as if she had never laid it in the first place. This behavior varied only slightly from bird to bird. The ringed turtle dove, in contrast, was so tame she suffered the indignity of having her eggs removed without moving. After they were gone, she stirred a bit and, realizing something was amiss, lowered her head to get a better view. Looking up, she spied the eggs and either sat there awhile mulling over the turn of events before concluding she had done all she could or trying to retrieve one egg. If the retrieval proved successful, she once again resumed incubating without concern for the other egg. But there was far more variation in how the individual turtle doves responded than among the passenger pigeons. The third species tested was the rock pigeon, what Whitman called the dovecot pigeon. The hens of this species will seek to regain the two eggs and, if they fail, will abandon the nest after greater hesitation than either of their cousins.[12]

Although Whitman might have had a special place in his heart for the passenger pigeons, his conclusion from this experiment would certainly not sit well with the subjects had they been human: "The passenger pigeon's instinct is wound up to a high point of uniformity and promptness, and their conduct is almost too blindly regular to be credited with even that stupidity which implies a grain of intelligence." Whitman, in making this statement, seems to have overlooked two relevant facts. First, the ringed turtle dove is a species of human creation, being a mix of many strains bred over generations. Similarly, most varieties of rock pigeon are also the product of human agency. Second, since passenger pigeons nested in trees, a missing egg would have been irretrievable. But that was not necessarily true for rock pigeons, which nested on ledges: the egg may merely have rolled a short distance away.[13]

This flock of passenger pigeons would be the only one ever studied by scientists. Most of the attention devoted to the birds came from Wallace Craig, an associate whose principal interest was in animal behavior and whose observations of the pigeon are incorporated into earlier chapters of this book. Without his efforts, the gaps in what is known about this species would be even larger. Craig seemed to have genuine affection for the bird and regrets that it was not more widely kept: "As an aviary bird, it would have been a favorite, on account of its beauty and its marked individuality . . . And for such study at close range the Passenger Pigeon was, and would ever have continued to be, a most interesting subject, for its strongly marked character appeared in every minute detail of its habits, postures, gestures, and voice . . . Such individuality is in great part impossible to describe, though it is felt unmistakably by everyone who has lived with the birds."[14]

As director of the Marine Biological Laboratory, Whitman considered it essential that he spend his summers at Woods Hole. So twice a summer, he carefully packed his hundreds of pigeons on either a "poorly ventilated baggage car" or "two freight cars" for the three-day trip between Illinois and Massachusetts. The irony is difficult to miss: these passenger pigeons never felt the exhilaration or weariness of extended flight for the only migrations they ever experienced were via train, a conveyance that played an important role in transporting the species to oblivion.[15]

From the high number of sixteen at the beginning of 1902, the flock began an irreversible downward spiral. The fate of Whitman's collection might be seen as a microcosm of the entire species. Eggs were laid, a few hatched, but not one young bird survived. Over the same time, the adults began to disappear: two escaped (they flew the coop at Woods Hole), two fell victim to tuberculosis, and others succumbed to causes unknown. One hen was given to the Cincinnati Zoo in 1902. Five years later, tuberculosis claimed the last of Whitman's birds, a pair of females.[16]

Although Whitman expressed his desire to have perpetuated the passenger pigeons under his care, conservation seemed of little interest to him. In this, he was typical of the majority of academics and indeed his countrymen. Referring to passenger pigeons, the late historian Philip Pauly writes, "The fraying of the thread connecting American academic biologists to the continent, its organisms, and their problems can be seen, however, in the poignant life of one of the more unusual summer migrants to Woods Hole." Several more decades would have to transpire before a formal push began to enlist academicians in promoting the preservation of biodiversity. But among the advocates were a number who had been Whitman's students.[17]

Whitman dutifully answered letters about his passenger pigeons and was generous in providing photographs of them for use in numerous publications. I would say that over 90 percent of all photos of live passenger pigeons were of his birds. But he never published anything on their plight. Even worse, particularly given how conscientious he was in his own research, he failed to read the available literature that would have enlightened him that the pigeons did eat animal matter. The untrained Whittaker knew to feed his breeding birds worms and insects. When Whitman finally made the discovery on his own, he was sorry for not having furnished additional protein to his birds earlier, for they would probably have been healthier and produced more young. Pauly also suggests that some of the trips east might have coincided with the breeding period, thus further inhibiting reproduction.[18]

Whitman's flock was gone except for that one female he had sent to the Cincinnati Zoo.

CINCINNATI

In 1872 Cincinnati endured a major infestation of caterpillars that showed a great fondness for tree leaves. Concerned over the future of the city's arboreal foliage, starch magnate Andrew Erkenbrecher helped found the Society for the Acclimatization of Birds. Under its auspices, a member departed for Europe with instructions to identify and bring back those insectivorous birds that might help stanch the outbreak. The society bought cages to hold the expected arrivals, which numbered about a thousand, consisting of such species as Eurasian starlings, house sparrows, and nightingales. A majority of them were liberated in May of the following year. The nightingales and starlings couldn't acclimatize and died out. But subsequent releases in various parts of the country ensured that the starlings and house sparrows, at least, continue to this day, carrying on their valiant struggle against insects, seeds, and the nestlings of native birds.[19]

Erkenbrecher had a long-standing love for animals and admired the zoos of his native Germany. He had the Acclimatization Society write to Alfred Brehm, former director of the Hamburg zoo, for "information on establishing a zoological garden." Brehm's positive reply inspired the creation of the Zoological Society of Cincinnati in July 1873. Under Erkenbrecher's continued guidance, this new society raised funds, bought property, commissioned plans, erected buildings, moved earth, laid paths, hired staff, and acquired animals.[20]

Despite all the challenges inherent in such an ambitious undertaking, the Cincinnati Zoological Gardens, now formally named the Cincinnati Zoo & Botanical Garden, commenced serving the public on September 18, 1875, thus becoming the nation's second-oldest zoo. (Philadelphia's zoo opened the previous year.) The eclectic assortment of 259 mammals included four species of bears, thirteen species of monkeys, six raccoons, two elks, a wombat, fifty-eight prairie dogs, a blind hyena, and a circus elephant accused of being ill behaved. The zoo also boasted two alligators and fourteen other reptiles. But mostly, the collection consisted of 494 birds, including a talking crow and a flock of peafowl.[21]

Passenger pigeons were part of the zoo's holdings from early on, and their last one died on September 1, 1914. That is known with certainty, but further details on the history of the species at the zoo are, in Schorger's assessment, hopelessly confused. For any given account, there is apt to be at least one other that contradicts it. What follows, then, is my best attempt at telling the story. Others have told it differently.[22]

A year before the zoo opened, a patron gave two passenger pigeons to the zoological society. The zoo may have held twenty-two of the birds when it opened. Whether they were actually on display at the zoo opening is not known for sure, although it seems likely they were. The zoo then added three more pairs in 1877 at the cost of $7.50 for the lot. Soon the birds became amorous, and early in March of either 1878 or 1879, zoo director Frank Thompson observed his wards mating: "I wove three rough platforms and fastened them up in convenient places; at the same time throwing a further supply of building material on the floor. Within twenty-four hours two of the platforms were selected . . . A single egg was soon laid in each nest and incubation commenced." Despite temperatures as low as fourteen degrees and the fall of so much snow that the nesting birds became completely covered, both the eggs hatched and the young fledged. Thompson wrote in an 1881 article that the flock stood at twenty despite the death of one old-timer.[23]

The flock inhabited a cage ten by twelve feet. Keepers provided a diet of cracked corn, wheat, cooked liver, and eggs. In contrast to the birds held by Whittaker and Whitman, which retained their wariness in the presence of people, the Cincinnati pigeons "became very tame, and when the keeper entered the cage . . . they would frequently alight on his shoulder."[24]

Martha's early time at the zoo is impossible to trace with any great confidence. This has plagued everyone who has tried to document her life. Most of the problem here arises from Salvator "Sol" Stephans, a former circus-elephant handler who took over the affairs of the zoo in 1878 and would stay for nearly half a century, until his retirement at the age of eighty-eight. He did a masterful job building the zoo into one of the country's finest, but he seemed to have little concern for consistency in answering questions about Martha. At times he said she hatched in the zoo. In 1907 he stated that the zoo's last female had been the bird conveyed by Whitman five years earlier. Other pigeons in the collection might have been called Martha, for Stephans told one writer that the bird, in this account born in the zoo, was named for the wife of a friend. But since her cage-mate was named George, it has generally been accepted that their monikers derived from the Washingtons.[25]

And don't even ask when the last surviving passenger pigeon was born.

Christopher Cokinos wryly handles the difficulty: "Sometime in 1902, 1900, 1897, 1896, 1895, 1894, 1889, 1888, 1887, 1886, or 1885, inside an egg . . . a female passenger pigeon tucked her bill between her body and a wing." It should be noted that Whitman probably did not keep track of when a given adult had been born either, for nowhere does he give the age of the bird that he sent to the Cincinnati Zoo.[26]

Martha settled in. She fed, she rested, and she fluttered a bit. In her younger days, she might have tried her bill at nest building. It is possible she laid some infertile eggs. As the passing years took their toll, she watched the members of the flock slowly disappear. Stephans tried to augment the flock, but no pigeons were to be had. It is easy to become anthropomorphic about Martha's situation as the idea of impending aloneness so absolute is heart-rending, especially in light of what had been such a short time before.

Meanwhile, by 1907 the Milwaukee pigeons had dwindled to four old males thought to be well beyond breeding age. Trying to bring the two remaining flocks together at that late date, therefore, would not have helped prolong the species, even if the idea had crossed anyone's mind. Whittaker had relinquished his birds to another, who apparently kept them in squalid conditions. These last stalwarts all supposedly succumbed to tuberculosis between November 1908 and February 1909, and their bodies were discarded because they were deemed in poor plumage due to a delayed molt. Knowing how valuable these specimens were alive, their keepers still treated them as garbage in death. Collectors would surely have paid something for them, or at the very least, a museum somewhere would have accepted them, even if they did so without the enthusiasm they would have shown for pretty corpses.[27]

The year 1909 was a tough one for the Cincinnati birds, too. One old cock died in February. Advanced age was thought to have killed him, but he, too, was in molt and thus also thrown away. When news spread that Cincinnati had the only extant pair of passenger pigeons, Stephans reported that other zoos offered up to $1,000 for Martha and George. George would last one more year, eventually as weakening that he had trouble walking. His infirmity ended on July 10, 1910 and, as far as I know, he was thrown out as well.[28]

Martha had sole reign of her environs for another four years or so. Her fame grew and people made her aviary a destination. The New York Zoo is said to have done all they could to get Stephans to part with his unique exhibit. Protected from the violence that would have claimed her in nature, Martha's vitality slowly ebbed. Keepers lowered her perch so it was mere inches above the floor. She rarely moved anymore, hardly the performance expected by the crowds. Joseph Stephans, Sol's son, said that "on Sundays we would rope off the cage to keep the public from throwing sand at her to make her walk around."[29]

A reporter gave this description of Martha: "There will be no mistaking the bird, as its drooping wings, atremble with the palsy of extreme old age, and the white feathers in the tail, make [her] a conspicuous object." Nothing beautiful, rousing, or frightening to see here, folks. Just an old bird near the end of life, almost agonal. Move along.[30]

Due to Stephans, even the details of her death are in dispute. Most accounts say she died on September 1, 1914. Almost certainly, this was sometime after the noon hour, most probably closer to one, although it might have been four hours later. Keeper William Bruntz might have discovered her crumpled body, or perhaps the Stephanses kept her company as the life force reluctantly flickered to its conclusion, bringing closure to the feathered whirlwind that defied human understanding, if not the human capacity to destroy.[31]

Martha was in molt when she died, but this time feathers were saved so they could be reattached to make her mount look more comely. She was, after all, headed to the Smithsonian Institution, and a specimen must look its best when it is slated to spend eternity as an exhibit in the National Museum. But to make sure Martha would weather the trip east, Joseph Stephans took her body to the Cincinnati Ice Company, where she was frozen in a three-hundred-pound cube of ice. Three days on a train in extreme heat reduced the frozen chunk substantially, perhaps completely, as one article says she reached her destination in a puddle of meltwater. But a photo exists of Martha lying on a slab of pitted ice that was most likely taken at the Smithsonian where she arrived on September 4.[32]

The Smithsonian fully understood the importance of its new specimen. William Palmer and physician R. W. Shufeldt were quickly dispatched to photograph, skin, and necropsy the bird. To Shufeldt, Martha "had the appearance of a specimen in health, with healthy eyes, eye-lids, nostrils, and mouthparts. The feet were of a deep, flesh-colored pink, clean and healthy, while the claws presented no evidences indicative of unusual age." They photographed the bird in several poses, then Shufeldt continued taking pictures as Palmer skinned her.[33]

At the end of their first day's work, Palmer took the skin so it could be stuffed and mounted by the taxidermy department, while Shufeldt performed a minute examination of the bird's internal organs. He noted the "great size" of the pectoral muscles, endowing the species with its remarkable powers of flight. More surprising was the discovery of a tiny slit in the right side of the abdomen "from which blood was oozing." Upon his enlarging this opening, he saw that the "right lobe of the liver and the intestine almost entirely broken up, as though it had been done with some instrument. As to the intestine, it was missing altogether." No explanation was offered, and one more mystery attaches to this iconic bird shrouded in so many.[34]

I was disappointed to learn that in addition to his valuable work in avian osteology and paleontology, Shufeldt authored a vile screed on domestic race relations. So while he had no regard for many of his fellow citizens, he was moved by the object on his dissecting table. When he reached Martha's heart, his own took over: "There is every reason to believe that the internal anatomy of . . . this heart of the Passenger Pigeon agree, in all structural particulars, with the corresponding ones in any large wild pigeon . . . I therefore did not further dissect the heart, preferring to preserve it in its entirety, —perhaps somewhat influenced by sentimental reasons, as the heart of the last 'Blue Pigeon' that the world will ever see alive."[35]

CHAPTER 10

Extinction and Beyond

All thinking people now realize that man alone was responsible for the extinction of the passenger pigeon so that further discussion of this phase of the subject is unnecessary.

—A. W. SCHORGER

Many had difficulty acknowledging that the passenger pigeon had reached extinction. Such a thing was simply not possible. There had to be some other explanation for their apparent absence. *Science* published an article whose author claimed that the species had retreated in great numbers to the deserts of Arizona. A *Condor* reviewer minced no words in responding, pointing out that since the writer had confused such distinctive birds as California quail and Gambel's quail, there was no reason to believe he could identify passenger pigeons. The *Auk* ran a note suggesting that the birds had escaped "persecution" by hiding out in an "extensive plain" fifteen miles east of Puget Sound. The report dated from 1877, but the author seems to endorse the idea that the entire population of the species might be moving around in places it was never known to occur: "Every bird lover would rejoice to hear that this wonderful bird had finally outwitted its great persecutor and lengthened its lease on life by 'going West' in the true American spirit of liberty."[1]

At least nothing in these musings suggested that the birds had abandoned the continent, unlike the Wisconsin lumberman familiar with the species who "saw millions of the genuine old time Passenger Pigeons" in the pine forests of Chile. Others placed the birds in Australia. But it wasn't enough for some that the crafty pigeons were hiding out in remote locations where no one would think to look. For them, the real reason the pigeon vanished is that it evolved into a different form whose behavior and appearance rendered it unrecognizable to ornithologists. Under this scenario the species had assumed a disguise, a renovated plumage far more "gorgeous" than before, and more befitting of its new tropical home in Colombian jungles.[2]

Acceptance of the pigeon's extinction came easier to a different group of unorthodox thinkers. Yes, tragically, there were no more pigeons, but what in the world could have caused their demise? Many of these people held that the birds had all drowned while crossing a large body of water, usually one of the Great Lakes or the Gulf of Mexico. Henry Ford, according to his secretary in a letter to W. B. Mershon, evidently believed the harried birds perished in the Pacific Ocean as they dashed to freedom in Asia. Sometimes the watery deaths were not occasioned by volitional acts of the bird but by powerful windstorms or dense fog that confused the mighty flocks into veering far offshore.[3]

More plausible notions have been advanced to explain, at least in part, what happened. The most persistent of these is that disease thinned the feathered hordes. It has the appeal of minimizing the human influence and allows for a speedy demise. Regrettably, for proponents, no evidence for this exists. Nothing in the literature points to masses of birds dying for unknown reasons, and most modern commentators have rejected this.[4]

In recent times, other efforts have been made to downplay the central role that direct human exploitation played in the species' extinction. It is almost as if sheer slaughter is too simplistic and more intricate ecological factors must be at play. One author of a prominent article in 1992 had evidently combed the literature looking for novel explanations. So there was the assertion that birds required a nesting area with recently melted snow to provide ample ground forage. This ignores the nestings in Mississippi, Oklahoma, Kentucky, and many northern locations where there was no heavy snow the year before. Such a situation also prevailed in the Petoskey area in 1878. This author also attributed competition for acorns with feral swine as another important factor in the bird's demise. The example of Texas vitiates this argument. The state supported large numbers of feral pigs, and the passenger pigeons seemed to do fine despite their porcine competitors. The pigeons were so persistent, which surely they would not have been had they found insufficient forage, the locals resorted to extreme measures in driving them out, including burning the woods that supported both the birds and the pigs. To support the notion that regardless of the slaughter only habitat loss could reduce the number of pigeons, Audubon's opinion is cited; but Audubon made the statement while the species was still abundant, and it was based in part on his false beliefs that passenger pigeons laid two eggs and nested multiple times during the year.[5]

In assessing the reasons for the bird's extinction, some things have been overlooked. Most accounts say that the decline was not sudden but took place over three centuries. They quote Josselyn (1674), Kalm (1750), and Smith (1765) to the effect that people had noticed a decline in the species for a long time. As the East Coast and other areas became settled, they obviously became less able to host the same number of birds that they used to. By 1840 or so, "great nestings became few and far between in the east," wrote Forbush. No doubt the population was less than it was at its highest point, whenever that was.[6]

But Major King's amazing flight of over a billion birds took place in 1860. Schorger estimates that eleven years later, the Wisconsin Dells nesting, which in his view comprised almost the entire population then existing, numbered 136 million nesting birds (this omitted non-nesting birds and squabs). The additional pigeons that nested in Minnesota that same time raises the total by several million. There was also a reported nesting that year in Cochrane County, Ontario, said to be in the millions. So if King was remotely accurate in describing what he saw, and Schorger was in the ballpark in his estimate of the Wisconsin nesting (and even adding millions more birds from Minnesota and Ontario), then a reduction of no less than a billion birds occurred in eleven years. If there were a lot more birds in 1871 than Schorger estimated, the window of collapse would be extended a decade. And even if King's description is wildly exaggerated, the population went from hundreds of millions to zero in only forty years.

Another reality that is often overlooked in trying to fathom why the species' population plummeted in decades is that little about the bird's requirements are known with certainty. What is indisputable is that their population was bewilderingly vast in 1860 but virtually gone by 1900. As the preceding chapters also demonstrate, *Homo sapiens* slaughtered the bird methodically and relentlessly. Most everything else is a matter of speculation.

For example, although sometimes overlooked by superficial examinations of the record, the birds did not always nest in the gargantuan concentrations that received almost all of the attention. Nestings ranged from a pair or two through small colonies to large colonies. The assumption is always that these attempts rarely succeeded, what with a single egg, conspicuous nest, etc., but the data are nonexistent, for no one allowed the birds to live long enough to find out, or if some amazing oddball did actually have the patience and self-restraint to check, the findings have been lost or have at least eluded me. But although such observations are lacking, at least some of these nesting efforts did result in fledged young because a number of the last specimens, including Buttons herself, were of first-year birds.[7]

There is also the assumption that the birds could not survive in the absence of mast. Mast was their preference during nesting, but they ate lots of things, and vast flocks often relied on other foods over varying periods at other times of the year. The rich nuts were highly nutritious, but that birds could have been sustained during a given nesting season by another food source seems likely, even if the number of young produced was possibly lessened. They nested well north of oak lands in Ontario, and an early account details a large nesting in the vicinity of Moose Factory, at the southern end of James Bay. Excluding oak, beech, and chestnuts, Schorger lists close to forty genera of wild plants known to have been eaten by passenger pigeons. In addition, they preyed on earthworms, snails, and numerous types of insects.[8]

If, as the above paragraph argues, the birds were indeed more catholic in their tastes, it means they had fewer restrictions on where they could live and nest. But even if that was not the case, more than enough nut trees remained to keep the birds going well beyond the 1880s. The first trees to be cut in northern areas were pine, and high-quality pigeon forage such as oak (particularly), birch, and aspen often appeared soon thereafter to colonize open ground. Eventually the accelerating loss of forest habitat would almost certainly have reduced the bird's population beyond the point of sustainability (at least in the absence of significant human assistance), but the species did not last long enough for that to happen.[9]

So, then, what did happen to the passenger pigeon? In the words of filmmaker David Mrazek, "You could say we happened to the pigeon." Europeans began the killing sometime during the first five months of 1565 (at least that is the earliest record I know of), and their successors, the residents of Canada and the United States, did not stop until there were literally no more birds left. When that happened, they shot mourning doves in the belief they were passenger pigeons. Virtually every time *Homo sapiens* crossed paths with the pigeons within killable distance, pigeons died.[10]

The intensity and thoroughness of the slaughter increased with the demographic and technological changes of the nineteenth century. The burgeoning populations of our urban centers needed cheap food. The wild game of the continent suffered inordinately to meet that demand, becoming a commodity fueling an industry. Since passenger pigeons were the most abundant dietary items among terrestrial vertebrates, few if any sources of high-quality protein cost less. They did not inhabit inaccessible parts of the west, nor were they sedentary creatures ensconced in inhospitable swamps far from probing human eyes. Their huge colonies could be reached. And whereas once the highly vagile species might have put down in some out-of-the-way place where they could find respite for a while, by the mid-1800s the growing ranks of telegraph operators made sure that the news of their whereabouts was soon humming through the lines to alert the hundreds of professional pigeoners and other interested parties. Remoteville, a hamlet tucked into the Missouri Ozarks, could quickly share news with the world.

Once the pros and local talent converged on the birds, the proximity of rail lines made transferral of the pigeons to larger regional and national markets easier and cheaper. As the 1881 Oklahoma nesting demonstrated, schlepping wagonloads of pigeons for three days across mountains to reach a train station failed to deter the dedicated pigeon dealer. At most, the difficulty in accessing rail transport merely reduced profits a bit.

As I have already indicated, habitat loss reduced the territory where the birds could collect in large numbers, but there was still plenty of food to sate the legions of pigeons. This whittling away of habitat did, however, make the birds easier to discover and kill. By the 1870s, most of the big nestings took place in Wisconsin, Minnesota, Michigan, Pennsylvania, and Ontario. Another idea that has been advanced is that as the birds decreased along with a more fragmented landscape, it may have been more difficult to locate the best places to roost and nest. In other words, fewer eyes were looking for suitable locations. On the other hand, there were also fewer stomachs to fill, so this factor is likely of limited importance, but may have had some marginal effect.[11]

A widely held view is that this species could not sustain itself without a large population. In part, it is the converse of why, presumably, passenger pigeons evolved the way they did: by massing in vast quantities they satiated any predator well before any serious harm could be done to the majority. This is the reason oaks and beeches produce huge crops in some years, and why periodic cicadas emerge in untold numbers every thirteen or seventeen years: the surfeit of nuts and insects ensures sufficient reproduction to keep the species going. As passenger pigeon numbers declined, they reached some threshold, still large for most species, below which they lost the capacity to make up for the high mortality they suffered through anthropogenic and other causes. The decline itself likely fostered increased mortality, making individual birds more vulnerable to predators. As one final nail, the ever-shrinking numbers may have lessened the stimulation needed for the synchrony of breeding that marked the big colonies.[12]

I have been immersed in the passenger pigeon literature since August 2009 and have devoted little time to anything else. That this spectacular and horrific extinction happened is clear; nor is there doubt that ceaseless, unbridled slaughter by human beings caused it. But I have struggled in accepting as sufficient the purported factors that reduced a billion or more birds to zero in four decades. This suddenness led to the fanciful explanations advanced during the early twentieth century and has long left me and others feeling unsatisfied.

But it becomes comprehensible when one thinks in terms of all that did and did not happen. For an avian species to sustain itself, a sufficient number of adults must lay fertile eggs, the eggs must hatch, the chicks must fledge, and the young birds must mature to begin the cycle over again: all the links in the chain of life must remain intact or the process fails. Most commentators have tended to focus on one or another of those links when in reality all of the links were simultaneously being compromised. The killing of the adults needs no elaboration. I have already noted that by the 1870s, the birds became more nervous and abandoned nesting sites with greater frequency. This may also have been a function of smaller numbers: the birds perhaps required the comfort afforded by having millions of neighbors. But with the ever-increasing intensity of exploitation, the birds became less tolerant of disturbance and quicker to give up their breeding aspirations.

While reduced, not all nesting terminated, however, and chicks were produced as evidenced by the barrels of squabs carted out of the big nestings throughout the 1870s. What has not received much attention—indeed only Derek Goodwin addresses it—is the purported practice of adult passenger pigeons, alone among all pigeons, to abandon their chubby chicks several days before they could fly. No pigeon species routinely abandons their flightless young, but most have low tolerance for molestation and are easily flushed off nests. A vast majority of these new chicks would have been doomed. The young birds seen rising in clouds from the colonies to join their parents would have referred to youngsters left alone much closer to fledging. With millions of birds present, dead chicks everywhere, and no one paying any attention to such nuances, the distinction between abandoned squabs whose flight was imminent versus those that were days away would have been lost. When the older young did fledge and took off to join the adults, it would have been reasonable to figure that the flock included the younger birds as well.[13]

Given the capacity to lay but one egg per annum, two years would have to elapse to replace the loss of every adult pair. With insufficient recruitment and nonstop killing that became even less restrained as the birds decreased, senescence caught up to the surviving adults and the species sputtered to an end. We may never fully understand or perhaps even identify all of the synergies that eventually led to 1914, for the historical record leaves unanswered certain questions. It is possible that some of these gaps in our knowledge may yet be filled through genetic analysis, which is being undertaken at several labs in the United States and Canada. But we should seek no solace in the presence of any contributory factors that may have come into play during those final decades of the bird's existence: if a person is shoved off a pier and drowns because he can't swim, culpability and the ultimate cause remain with the one who pushed.

THE ECHOES OF THEIR WINGS

> It is late. It is too late as to the wild pigeon. The buffalo is almost a thing of the past, but there still remains much to preserve, and we must act earnestly if we would accomplish good things.
>
> —REPRESENTATIVE JOHN FLETCHER LACEY, ON THE FLOOR OF THE HOUSE OF REPRESENTATIVES ON APRIL 30, 1900, INTRODUCING WHAT WOULD BECOME THE FIRST FEDERAL BIRD-PROTECTION LAW

Silence clutched the land and could not be ignored. No longer would the hum of pigeon wings rain down from vernal skies. The lowing of bison emanated from only a handful of animals, restricted to one vast tract of wilderness in northeastern Alberta and another flanking the headwaters of the Yellowstone River. Empty leks became mere reminders of the haunting strains of prairie-chicken music that formerly permeated the creeping dawn. Squawking rookeries of herons, gulls, and terns were stilled as their former inhabitants were transformed into hat decorations. Even the sharp reports of gunfire ebbed in many places as the legions of waterfowl and shorebirds that used to fill wagons were reduced to tiny flocks, if encountered at all. One needed to be neither perceptive nor learned to notice the change, for the evidence was everywhere.

Out of this quiet surged the country's first environmental movement. Three principal groups drove the concern that culminated in those early wildlife laws that still provide the basis for how the United States and Canada manage their biological heritage. The "scientific naturalists" provided the hard data and the organizational skills to coordinate the activities of their two major partners, who had very different perspectives from one another. The sportsmen's associations enjoyed large and influential memberships that were critical in furthering the conservation agenda. These associations were a mixed bag, however. The New York association, for example, put on shooting contests in which tens of thousands of passenger pigeons were killed, but had a solid track record in other aspects of conservation such as protecting fish stocks and preserving game. The final group is least well documented, for it consisted mostly of unheralded women and even children who wrote letters, collected money, and joined state Audubon societies in large numbers. They were moved largely by a distaste for cruelty and the gratuitous killing of animals.[14]

The American Ornithologists' Union (AOU) came into being in 1883 and produced a model bird law three years later. The proposed statute banned the killing of nongame birds, defined which birds were game birds, and made it unlawful to destroy the nests and eggs of all species, whichever of the two categories they were in. House sparrows, newly introduced from Europe, were specifically exempted from any mercy. Most of the work to persuade states to enact protective laws fell to the National Association of Audubon Societies, led by William Dutcher. Their efforts reaped great success, for over the next fifteen years, comprehensive bird-protection measures were ratified in fifteen states, thirteen of which adopted the model law. Only Idaho and the Indian Territories lacked any restrictions on the killing of nongame birds. Other states offered varying degrees of protection, but none more subjectively than Nebraska. After listing as protected such types as robin, "thrush," and "yellow bird," the Nebraska statute adds "or other bird or birds of like nature that promote agriculture and horticulture by feeding on noxious worms and insects, or that are attractive in appearance or cheerful in song."[15]

Connecticut not only provided complete protection for nongame birds but also prohibited the exportation of game birds legally obtained within its borders. This was an attempt to shut down the market hunting that had been so instrumental in eroding wildlife populations, which included of course the extinction of the passenger pigeon. But the wildlife industry rose to protect itself by claiming such a measure improperly poached in an exclusively federal domain by violating the constitutional edict that it is up to Congress "to regulate Commerce with foreign Nations, and among the several States, and with the Indian Tribes" (the "Commerce Clause").

The argument reached the U.S. Supreme Court. In a landmark decision, Justice Edward White wrote the majority opinion in *Geer v. Connecticut* (1896), which upheld the state measure: states held wildlife in trust for their people and had the power to ensure that harvested game was kept for their benefit; since this state power adhered to all steps in killing and distribution, full private ownership of the game was never attained, thus eliminating this "essential attribute of commerce." Even though the court clearly stated that whatever rights the states hold in their wildlife must be compatible with "the rights conveyed to the federal government by the Constitution," the decision gave ammunition to those who later claimed that any national laws abridging what states could do with their own wildlife were a breach of the commerce clause.[16]

But the federal government was extending its involvement in the promotion of conservation. The U.S. Commission on Fish and Fisheries was founded in 1871 to address depleted fisheries. President Theodore Roosevelt established the first national wildlife refuge in 1903. The first regulatory statute, though, was enacted in 1900. It became known as the Lacey Act, in honor of the Iowa congressman who toiled for years on behalf of bird protection. The law targeted the trade in wildlife that originated in states where its killing was legal but was exported into states where it would have been illegal. Under Lacey the legal status of all game entering a state would be based on the laws of the receiving state. Unfortunately, even with the Lacey Act, as long as some states refrained from passing adequate restrictions, interstate wildlife marketing would continue.[17]

The recognition that only the national government could provide effective relief for migrating animals, here today and gone tomorrow, led to the passage of the Weeks-McLean Act. Pioneering bird photographer and Pennsylvania congressman George Shiras III first introduced the legislation in 1904, but resistance to the idea that migratory wildlife was within the purview of the national government delayed enactment until 1913. Two federal district courts found it unconstitutional, but before the Supreme Court could weigh in, the United States and Great Britain, on behalf of its possession Canada, signed a treaty that protected the birds migrating across the national borders. (The United States would later sign similar agreements with Japan, Mexico, and the former Soviet Union.) To ratify the treaty, Congress passed the Migratory Bird Treaty Act, which was virtually identical to

Weeks-McLean. The high court dismissed the case against Weeks-McLean but soon visited the issue in *Missouri v. Holland* (1920), which challenged the new law as an improper usurpation of states' rights. Justice Oliver Wendell Holmes wrote a characteristically memorable decision: "To put the claim of the State upon title [to birds] is to lean upon a slender reed. Wild birds are not in the possession of anyone . . . But for the treaty and the statute there soon might be no bird for any powers to deal with."[18]

One other early law is worth noting. In 1913 the Wilson Tariff Act stopped any further importation of wild-bird feathers for hats or other business uses. Along with Lacey, it was subsequently changed to ban the importation of wildlife that was obtained or exported illegally from the country of origin. These amendments represent the first time the laws of one country were used to help the effectiveness of another. Both laws provided the underpinnings of the U.S. Endangered Species Act (1973) and the Convention on International Trade in Endangered Species of Wild Fauna and Flora (CITES), the principal legal bulwarks in the nation's efforts to protect and recover imperiled plants and animals.

Congressman Lacey's words from 1900 ring loud and clear.

With regulated hunting and the establishment of public refuges and preserves at all levels of government, wildlife in general increased dramatically, and certain species on the precipice of extinction moved to safer ground. If the passenger pigeon is the icon of an animal driven to extinction through deliberate, wanton, and direct human actions, the continued existence of other species proves that these new conservation measures were effective. Bison now roam throughout the west and have become agricultural products as privately held herds are managed like cattle for meat production. Great egrets, plundered for their nuptial plumes and restricted to a few isolated or protected rookeries in the Deep South, rebounded and now occupy their former territory throughout the eastern United States. The heaviest of North American birds, the trumpeter swan, retreated before the gun and dredge to make its last stand at a few remote locations in Montana, Wyoming, Alaska, and western Canada. In 1933, the lower forty-eight states harbored only sixty-six of the swans, all in the Yellowstone area. Under close protection, however, their population is now pegged at sixteen thousand, enabling the species to reclaim lost range; augmented by introductions, trumpeter swans again inhabit parts of the Great Plains and Midwest after an absence of 150 years.[19]

As time passed, the technologies needed to sustain growing human populations took new forms. Other threats to biodiversity emerged or at least became serious enough that they could no longer be ignored. Rachel Carson's clarion call in *Silent Spring* alerted the populace that products whose last four letters are based on the Latin word for "kill" (*cide*) are often less selective than claimed by their manufacturers. The broader implications that modern industrial societies sullied the air and the water with poisons as a matter of course led to the second great environmental movement of the late 1960s and early 1970s.

Again laws were enacted and toughened. In the United States these included the Clean Air Act, Clean Water Act, National Environmental Policy Act, and the Endangered Species Act (1973). Canada passed the Environmental Protection Act and the Species at Risk Act, among others. The banning of certain pesticides in North America and the improvement in water quality have done wonders for such fish-eating birds as ospreys, bald eagles, double-crested cormorants, and white pelicans. White pelicans now nest farther east than the historical record indicates they ever did previously.

Actions under the Endangered Species Act have contributed to the "recovery" of twenty-two species or subspecies of plants and animals, including Eggert's sunflower (*Helianthus eggertii*), peregrine falcon, grizzly bear, most populations of gray wolf, and all but one population of bald eagle. Other organisms have experienced increases in population even if their long-term survival is still uncertain; black-footed ferret, Big Bend Gambusia (a small fish, *Gambusia gaigei*), whooping crane, California condor, and Kirtland's warbler are five examples.

The Kirtland's warbler is an unusual case. Its survival depends on government largesse. As far as anyone knows, it was always rare, for its geographic range is largely limited to the northern part of Michigan's Lower Peninsula (a small number of birds now nest in the Upper Peninsula, Wisconsin, and Ontario). Within that circumscribed area, the warbler breeds almost exclusively in jack pines (*Pinus banksiana*) between five and twenty feet in height. Add to those narrow habitat requirements heavy parasitism by brown-headed cowbirds, which routinely lay their eggs in the nests of the warblers, and you have a formula for extinction. But because of intensive cowbird control and habitat management costing over a million dollars a year, this distinctive bird has a future.

The mere passage of legislation does not, of course, prevent all extinctions. Take the dusky seaside sparrow (*Ammodramus maritimus nigrescens*), the only animal I have seen that became extinct. When I added the bird to my life list in June of 1972, it inhabited a paltry section of brackish marshland near Cape Kennedy, Florida. As a birder, I was disappointed when it was later relegated to the status of a subspecies of the much more widely distributed seaside sparrow. Fortunately it was not abandoned, although what efforts were expended on its behalf bore no fruit. But its story illustrates well how much the forces of extinction have changed since railcars filled with wild-pigeon carcasses rolled out of Petoskey.

You might say the dusky seaside sparrow was an unintended casualty of mosquito control, *collateral damage* in military parlance. Dousing of the bird's habitat with pesticides reduced its population substantially. No doubt the poison also aided in the control of the pesky insects, but apparently not enough to make the environs of the Kennedy Space Center comfortable to personnel. To move forward with the jihad against mosquitoes, a large part of the remaining sparrow range was flooded. Then, to widen a highway, the little bit of marsh that still held the few sparrows was drained. By 1979, the world population of dusky seasides numbered six, all males. Four years later, the four survivors were taken to the Disney World Resort, where attempts to breed them with another dark race of seaside sparrow failed. The end came on June 17, 1987, almost fifteen years to the day after I saw them. I have always thought that the last of a species should never expire at Disney World, but at least he was never put on display where the public could throw sand to make him move, an indignity to which Martha was subjected.

Now, as we approach the centenary of the passenger pigeon's extinction, most scientists agree that we are currently in the early to mid stages of the planet's sixth great episode of mass extinctions. (This is despite the laws and advances in knowledge.) Not only does this episode differ from previous ones in being caused by one species—some refer to human activities as the modern analogue of the asteroid that destroyed the dinosaurs—but it has resulted in the "loss of larger-bodied animals in general and apex species in particular." Apex species are predators that have no predators themselves and reside at the highest trophic levels (they inhabit the top of the food chain). Their elimination is especially pernicious because it can profoundly alter the ecological structure and dynamics of ecosystems, including such things as fire frequency, disease, invasive species, and biochemical processes.[20]

A report from the International Union for Conservation of Nature in 2010 says that 30 percent of amphibians are at risk of extinction; 21 percent of mammals, reptiles, and fish; and 12 percent of birds. The Alliance for Zero Extinction has identified 131 mammals, 23 coniferous trees, 15 reptiles, 217 birds, and 408 amphibians from across the world that face imminent extinction. Nature Canada lists 631 species facing extirpation in their country, while authorities in the United States officially recognize nearly 1,100 plants and animals facing the same fate in their nation. Slightly more secure are 295 species deemed to be only threatened.

Biologist David Blockstein refers to the Four Horseman of modern extinctions: habitat loss, direct take, pollution (which would include pollution-induced climate change), and introduced species. Each of these can be devastating to a particular group of species, but these factors are not operating independently. Often, an interplay between them exacerbates the effects of any one. It is important to keep this in mind as we discuss briefly the anthropogenic forces that are ridding the planet of its biological richness.[21]

Habitat loss is impossible to miss, for modern human societies consume tremendous amounts of space. Landscapes once harboring a multitude of living things are covered with buildings, roads, and croplands that ooze across the countryside with nearly the same lethality as molten lava. Mining and unsustainable forestry practices gobble up ground in the hinterlands. From the perspective of most plants and animals, humans make terrible neighbors. Relatively few organisms find these new circumstances hospitable and proliferate, often to the detriment of less common species. But many more find that these changes make it impossible to execute life's necessary functions. Habitat loss is generally deemed to pose the greatest risk to biodiversity.

One subset of habitat loss that carries its own special type of threat is the imposing structures that we design and construct. Tall buildings are monuments of prestige and wealth. But when well lit during periods of migration or veneered with reflective glass, they become deadly obstacle courses for birds and bats. One biologist concluded that only habitat loss claimed more avian lives than collisions with buildings. Communication towers potentially threaten air traffic, so those higher than two hundred feet or near airports must be lit, but that illumination, particularly when it is red, increases crashes of another kind: millions of birds, of which various warblers, including the rapidly declining cerulean, are among the principal victims. If the construction of wind turbines, an environmentally friendly energy source, continues to expand as projected, a million more birds may die per annum. Bats are not faring too well either, as their corpses have been found at almost every wind facility in North America where adequate studies have been conducted.[22]

While regulated hunting is unlikely to cause extinctions, illegal taking of species whose carcasses bring high prices continues to plague conservation efforts on behalf of such animals as the tiger and several species of rhinoceros. In the case of horseshoe crabs, overharvesting along the Mid-Atlantic coast of the United States has not only reduced the crabs but is damaging shorebirds, particularly the red knot, that rely on an abundance of crab eggs for food before embarking on long flights to arctic breeding grounds.[23]

But the most egregious examples of profligate slaughter, reminiscent of the type that so ravaged wildlife in the nineteenth century, occur where there is no state authority: the open ocean. Factory vessels equipped with immense seines can remove virtually all life larger than its meshes from a column of water that reaches from the surface to the bottom. Catches of the five most coveted species of tuna, for example, have increased from a million tons a year in the mid-1960s to over four million currently. These species are now deemed to be threatened or nearly threatened with extinction by the International Union for Conservation of Nature. But where there is no authority to stop them, players are free to plunder. Only international cooperation can spare the commons from becoming defiled and depauperate. Yet with the world population of people topping seven billion, that kind of voluntary restraint seems like a daydream.[24]

Pollution is a seemingly unavoidable concomitant of human aggregations that only increases with population size and standards of living. Specific contaminants can be regulated or banned, but the pernicious effects of spewing forth pollutants in huge quantities is felt across the globe. Aquatic ecosystems are perhaps most vulnerable as the introduction of pollutants renders the water unfit for life. This effluent can kill outright or negatively impact long-term health by altering reproductive capacity, creating specific illnesses such as cancer, reducing oxygen levels and water clarity, compromising immune systems, increasing levels of acidity . . . and the list continues. Even the oceans, the planet's least vulnerable repositories of liquid water, are being sullied to dangerous levels: they are beginning to manifest what appears to be impending biological impoverishment beyond what humans have ever experienced, and the time to implement effective and rapid remedies is fast ebbing.

As the industrial world has dumped untold tons of carbon into the air for centuries now, we have triggered yet another profound challenge to the maintenance of biodiversity. The effects of climate change are just beginning to be felt and are most pronounced at high elevations and in the Arctic, where the warming is apt to be greater than in other regions. Whitebark pines (*Pinus albicaulis*), a keystone species of subalpine forests, inhabit western Canada and the United States, but are fast disappearing as temperatures rise and moisture decreases. These changing conditions favor competing trees and make the pines more susceptible to fatal infestations of mountain pine beetle (*Dendroctonus ponderosae*) and white pine blister rust (*Cronartium ribicola*), a native of Europe. Fire suppression by land managers also allows expansion of competing trees at the expense of whitebark pines.[25]

The Arctic Climate Impact Assessment projects that warming will reduce summer sea ice by roughly half in less than a century. This will likely prove devastating to a host of animals that require ice for resting, feeding, or denning. Such charismatic species as polar bears, ribbon seals, walrus, and ivory gulls are among those that would be affected. One known victim of climate change is the terrestrial Peary caribou (*Rangifer tarandus pearyi*), an inhabitant of the high arctic islands of Canada. This herbivore survives by using its unusually pointed hooves to reach food plants buried beneath the snow. But freezing rains and a change in melt-thaw cycles, both caused by a greater number of warmer days, have led to the formation of ice layers that stymie efforts at foraging. This is believed to be a major reason behind the caribou's population crash: from twenty-six thousand in 1961 to seven hundred in 2009. Increases in ice and crusted snow have also adversely affected such other arctic land mammals as musk oxen, lemmings, and voles.[26]

Organisms evolve in specific ecosystems. Over millennia they adapt so as to most effectively deal with the physical conditions of their range and to coexist with the other life-forms with which they share space and compete for resources. They develop physical attributes and behaviors that allow them to procure food or avoid predators. Long-term exposure to pathogens means that surviving hosts build resistance. But when a stranger appears, new to the evolutionary history of the established fauna and flora, calamity often results.

Niagara Falls prevents Atlantic fish from ascending the Great Lakes any farther than the western end of Lake Ontario. The construction of canals, however, has allowed sea lampreys and alewives to circumvent the forbidding barrier and colonize the upper lakes, leaving biological desolation in their wake. Sea lampreys use their jawless mouths armed with tiny, sharp teeth to attach themselves to larger fish, whose bodily fluids provide sustenance to the new passengers. When the host is a saltwater species weighing hundreds of pounds, little harm ensues, but when the victim is a twenty-pound lake trout, the chances of its survival plummet. Alewives, on the other hand, feed on plankton, but reached such numbers they outcompeted the native planktors. Lake Michigan provides a stark example of how these two oceanic fish devastated the existing ecosystem. Burbot, a type of cod, and lake trout, for forty years the most valuable fishery in the lake, swam atop the deepwater food chain, preying on seven species of ciscoes or chubs that foraged on plankton. Already reduced by overfishing, the trout and all but one of the ciscoes proved highly vulnerable to the invaders and perished from the lake within five decades of the lamprey arrival. Four of the ciscoes no longer exist anywhere on the planet.[27]

The brown tree snake is native to Australasia. When it arrived in Guam as a stowaway, it wiped out the bird life of its new home, driving at least one species to extinction and another to virtual extinction. But nowhere has avian diversity suffered more than in the Hawaiian archipelago, which holds the distinction of having lost more of its native birds due to anthropogenic causes than any other place in the world: seventy-one endemic species have vanished since human beings first arrived. Much of this destruction was caused by mosquitoes, which were not present on Hawaii until their inadvertent introduction in 1826. These bloodsucking insects carry and spread avian malaria and avian pox virus, two deadly diseases also previously unknown to the local birds. Currently almost no native land birds are found at elevations lower than four thousand feet, the highest altitude mosquitoes can tolerate.[28]

North American birds took a big hit with the introduction of West Nile virus, a recent arrival from Africa and Eurasia that also infects birds through the feeding habits of mosquitoes. As the disease swept across the continent and then south into Mexico and South America, the newly affected landscapes supported many fewer crows, jays, magpies, hawks, and sage grouse than before. The yellow-billed magpie is restricted to portions of California, and in two years the virus reduced the entire population by almost half. But populations are recovering as the avian survivors develop immunity.[29]

Another disease currently running rampant through eastern North American is ravaging bat populations. White-nose syndrome, caused by the fungus *Geomyces destructans*, was first noted in a New York cave in 2006. Since then, close to 6 million bats of nine species have died. Three of the affected species—Indiana myotis, gray myotis, and Virginia big-eared bats—were already federally endangered. Professor Thomas Kunz of Boston University likened the magnitude of this population decline to the major subject of this book: "I think you have to go back to the 1800s, to the loss of the passenger pigeon, to find something similar."[30]

The syndrome irritates the skin, which causes bats to awake more frequently during hibernation. This additional activity depletes energy reserves they need to survive the winter. In some places, mortality rates of infested bats have reached 100 percent. Most likely, the fungus is a European native that was introduced to North American caves by human visitors. But it is also possible, however, that the fungus has always been native to these shores but has only recently mutated into the form that is causing so much harm.[31]

More adorable than any fungus, cats (*Felis catus*), either feral or those whose irresponsible owners allow them to roam outdoors, are no slouches in their depredations on native wildlife. These felines likely take in excess of 530 million birds a year in the United States. Small mammals, reptiles, and amphibians are on the menu as well. (And even those kitties subjected to clawectomies continue to be effective predators.) Birds restricted to islands have proven to be the most vulnerable, as cats have contributed to the extinction of thirty-three such species.[32]

Why should we care about extinction? In their classic 1981 book, *Extinction: The Causes and Consequences of the Disappearance of Species*, Paul and Anne Ehrlich list the four major categories of reasons why species should be preserved: (1) other life-forms have a right to exist, and ethical decisions should not be based solely on human benefits; (2) other species are aesthetically pleasing and add to human felicity by their beauty and character; (3) other species provide economic, medical, and other "direct benefits" by their continued existence; and (4) extinctions have indirect and long-term effects on ecosystems of which humans are also a part; the Ehrlichs illustrate this point with the analogy that the "popping of rivets" of an aircraft would eventually render the plane unable to fly. For sixty-five pages they elaborate on these reasons. I believe that all of these arguments are convincing, although valuing biodiversity is no longer an intellectual point for me. I take it as a given, a truth as primary as the Golden Rule or the efficacy of science.

Brian Anderson, director of the Illinois Natural History Survey, suggests that extinction be considered in personal terms. There is nothing esoteric or technical in this. Think of the excitement manifested on a child's face when the fishing pole she is holding begins reverberating with the power of a trout at the end of the line. The chorister hidden in the canopy of the neighbor's oak penetrates your consciousness and makes you wonder. Perhaps you savor the sights and scents of a late-April hike through the woods of a nearby park when the loam is blanketed with spring beauties, trillium, and Virginia bluebells. A full-antlered bull elk bugling against the shimmering gold of autumn aspen might forever be etched in your memory as a stunning tableau, or it might be inspiration for the coming bow season. And many are the gustatory pleasures associated with cod, lobster, and other morsels of the sea. How much poorer and less enjoyable would life be if these and other creatures vanished? But, of course, that is impossible.[33]

Appendix: A Passenger Pigeon Miscellany

I. CONSERVATION MEASURES: WAY TOO LITTLE, WAY TOO LATE

> Within four centuries of North American civilization (or modified barbarism) we can be credited with the wiping into the past of at least three species of animal life originally so phenomenally abundant and so strikingly characteristic in themselves as to evoke the wonders and amazement of the entire world.
>
> —GEORGE ATKINSON, "A REVIEW-HISTORY OF THE PASSENGER PIGEON IN MANITOBA," 1905

John Josselyn visited New England twice, the second time staying from 1663 to 1671. After describing the abundance of passenger pigeons (quoted in chapter 3), he is the first author who notes a decline in the bird's population: "But of late they are much diminished, the English taking them with Nets." A few other seventeenth- and eighteenth-century authors, including Kalm and Mather, collected similar sentiments. Kalm's informants thought the pigeon decline was due to a growing human population, the felling of timber, and increasing competition with swine for mast. These passages are significant not because they reflect an actual depletion in pigeon numbers on a range-wide scale, but because they are the first statements suggesting the possibility that the species *could* be vulnerable as a consequence of human behavior.

John Audubon seems to be the first to address the impacts that mass slaughter might have on the species' long-term prospects, and for that he deserves accolades even if he got it wrong. After describing the carnage at one huge roost, he tackles the obvious question: "Persons unacquainted with these birds might naturally conclude that such dreadful havoc would soon put an end to the species. But I have satisfied myself, by long observation, that nothing but the gradual diminution of our forests can accomplish their decrease, as they not infrequently quadruple their numbers yearly, and always at least double it." But in reaching this conclusion, he presumed incorrectly that the birds lay two eggs per nest and that they regularly nest multiple times a year.

As is so often the case, it took a foreigner to glimpse the truth. The de Tocqueville of passenger pigeons was the French writer Benedict Henry Revoil, who traveled through the United States in the 1840s. After having witnessed butchery at a pigeon roost at Hartford, Kentucky, in the fall of 1847, he demonstrated singular perspicacity in his astonishing prediction: "Everything leads to the belief that the pigeons, which cannot endure isolation and are forced to flee or to change their manners of living according to the rate in which the territory of North America will be populated more and more by the European inflow, will simply end by disappearing from this continent, and, if the world does not end this before a century, I will wager with the first hunter coming that the amateur of ornithology will find no more wild pigeons, except those in the Museums of Natural History." Events would show him to have been overly optimistic by about fifty years.

As the decades of the nineteenth century reached their final two or three, the occasional public statement would be made admonishing those engaged in the killing. The *Mauston* (Wisconsin) *Star* printed this dispatch from its New Lisbon correspondent in May 25, 1882: "Some of our prominent businessmen are busily engaged . . . in destroying thousands of poor little helpless young pigeons not yet with feathers, and encouraging others in the wholesale murder, and all from a greedy desire to catch a few squabs to ship and sell for thirty cents per dozen. Money! Fie upon you for shame." The *American Field* published a plea on behalf of the species the very next month: "But this fact is patent and admits of no argument to the contrary: that unless the trapping and shooting of the wild pigeon is stopped during the period the birds are nesting, extermination must necessarily follow, and rapidly." To protect one nesting in Wisconsin, a petition was drawn up and signed by various citizens requesting that the governor send out the state militia to prevent further hunting. J. B. Oviatt, the Pennsylvania pigeoner, said of his fellow netters, "If they knew that the pigeon were decreasing, they didn't want it known. For years I said the pigeons were decreasing, and they [most netters] were afraid a law would be passed, which there should have been" (Scherer 42).

In fact, laws relating to passenger pigeons were on the books, many of which stayed in effect long after the birds were gone. In 1848, Massachusetts became the first state to enact legislation dealing specifically with passenger pigeons. Rising up to quell a grave injustice, the legislature passed a law to protect *netters.* Should any miscreant impede the activities of this class of worker by firing guns or otherwise scaring off pigeons, he would be subject to a fine, damages, or punishment as a trespasser.

Vermont's statute three years later actually aimed to provide some protection to the bird, albeit not much. The pigeon was considered a nongame species, and thus the citizenry was prohibited from destroying eggs or nests. A violation could bring a fine of a dollar, but apparently the law was rarely enforced (Young 248).

The Ohio Senate looked at the need to protect passenger pigeons and issued findings that are widely quoted in the passenger pigeon literature. The 1857 report is ranked by the Ohio Historical Society as the fifth most embarrassing moment in the state's history, behind such events as the burning of the Cuyahoga River and the discovery in 1953 that Congress had never formally adopted a resolution admitting the state to the union. The legislative committee wrote, "The passenger pigeon needs no protection. Wonderfully prolific, having the vast forests of the North as its breeding grounds, travelling hundreds of miles in search of food, it is here to-day, and elsewhere tomorrow, and no ordinary destruction can lessen them or be missed from the myriads that are yearly produced."

Canada's restrictions on the exploitation of passenger pigeons lagged behind those of the states. Under its Small Bird Act of 1887, Ontario specifically excluded the pigeon as being among the birds that could no longer "be at any time killed or molested." A revision of the statute ten years later seems to have finally granted the protection that might have helped if there had been any birds left to protect. Game laws enacted in Quebec in 1899 and Manitoba in 1891 failed to provide the species any relief. The bird's extinction saved lawmakers the trouble of having to pass any protective measures (Mitchell 147).

Most of the laws that were eventually put in place in the United States modeled themselves after the one passed in New York in 1861. No disturbances were allowed at the actual nesting sites, and there could be no gunfire within a mile. The distances varied between states. Pennsylvania would also require that out-of-state pigeoners buy a county license for $50 to ply their trade, but apparently no one was ever prosecuted under the law. Massachusetts in 1888 did pass a potentially effective law when it banned pigeon hunting from May to October, when what few pigeons still existed were most likely to be in the state.

Michigan was the only jurisdiction to eventually ban all killing of passenger pigeons. As mentioned earlier, this occurred in 1897, just a year before the last pigeons were to be recorded from the state. But Michigan's first law affecting pigeons was in 1869. Under this statute, they were considered neither songbirds nor game birds and warranted their own separate section. A distinction was made between places where the birds nested and where they roosted. At nesting places, it was illegal to shoot pigeons within one half mile, but at roosts the shooting ban applied only to the actual site. Nothing restricted where or when netting could occur. Six years later the law was amended to provide a tiny bit more protection: the distance from a nesting where one could now lawfully shoot pigeons was extended to a mile. There was also a purported limitation on netting that read, "No person . . . shall, with trap, snare, or net, or other manner, take or attempt to take, or kill or destroy, or attempt to destroy, any wild pigeon, at or within two miles of such nesting place at any time from the beginning of the nesting until after the last hatching of such nesting." Note the final eight words: once the last eggs hatched, market hunters could have free rein. Protection adhered to the netters rather than the pigeons. And so the law stood until its revision in 1897, when for the first and only time in U.S. history a state granted passenger pigeons complete protection. That the provision called for the reinstatement of regulated hunting for the species in 1907 was really moot even at the time of its enactment. (Full citations are in the bibliography.)

II. SCIENTIFIC SCRUTINY

Cloning

Claims of cloning have been made for a frog, carp, and maybe twenty species of mammals. In many cases, the "successful" clone lived but a short time due to defects that are almost inevitable when shortcutting the intricate steps that have developed over millions of years of evolution. No bird has ever been cloned, not even those that are economically important, although poultry have been genetically manipulated.

All successful clones have been from tissues taken from living or freshly killed animals. The freezing and thawing of such tissues damages cells and destroys their suitability for cloning unless specific cryoprotectants are used. Still, the prospect of bringing back to life an extinct bird and releasing it into the wild is exciting enough for scientists and others to give it serious consideration. The Long Now Foundation hosted a meeting at the Harvard Medical School to discuss that possibility in January 2012. Of various theoretical approaches, the one with the most promise involved the extraction of DNA from passenger pigeons and using that to create passenger pigeon traits in a band-tailed pigeon. A host of challenges were identified, such as:

1. At what point, if ever, does a genetically altered band-tailed pigeon become a passenger pigeon?
2. If a handful of passenger pigeons could be created for life in a zoo, is that even worth doing?
3. Are vast flocks an essential attribute of passenger pigeons?

Even if all the scientific problems inherent in producing live passenger pigeons could be mastered, a host of political and other challenges of perhaps even greater difficulty would remain in creating wild, reproducing populations of the species. As of June 2013, additional discussions are being held and genetic research is moving forward.

(Based on my attendance at the meeting referred to, as well as additional information provided by Jennifer Schmidt, associate professor of biological sciences, University of Illinois–Chicago, and Ben Novak, lead researcher for de-extinction with the Long Now Foundation's Revive and Restore Project.)

Collections

Paul Hahn (1875–1962) first learned of the passenger pigeon through an article he read when he was twelve years old and living in Württemberg, Germany. He saw his first mounted passenger pigeon in 1902, four years after he and his family had immigrated to Ontario. The sight had a profound impact on him: "I was struck by its beauty and saddened by the knowledge that no one would ever again see the magnificent flocks which once darkened the sky."

As a personal memorial to the species, Hahn devoted himself to combing attics, bars, cellars, and every other kind of human habitation to collect passenger pigeons. From 1918 to 1960 he acquired seventy specimens, all of which he conveyed to the Royal Ontario Museum (ROM). Largely through his efforts, as well as those of James Fleming and others, the museum has 152 passenger pigeons (including ten egg sets), more than any other institution in the world.

This is known with certainty because of Hahn's other great work. In 1957 he became curious as to how many passenger pigeon specimens were known to exist, and he attempted to find out. At the suggestion of James Baillie at ROM, Hahn also included in his search comparable information on the great auk, Carolina parakeet, Eskimo curlew, Labrador duck, ivory-billed woodpecker, and whooping crane. Over five years he contacted collections all over the world, receiving responses from over a thousand. His final tally of passenger pigeons came to 16 skeletons and 1,532 skins and mounts. The work was published in 1963 as *Where Is That Vanished Bird?* Not surprisingly, an overwhelming majority of these specimens are in the United States and Canada. Based on my knowledge, which is certainly incomplete, there are specimens in every province except Newfoundland/Labrador (and none in the three territories) and every state but Hawaii, Maryland, Montana, New Mexico, Oklahoma (Schorger includes a photo of a lovely male from the National Cowboy Hall of Fame in Oklahoma City, but it is no longer there), and Virginia. (Since the hardcover edition came out, I have learned of passenger pigeon specimens in Georgia, at the Georgia Capitol Museum in Atlanta, and Wyoming, currently in the possession of a private owner who is considering donating it to a public institution.) It seems likely that collections within at least some of these states have passenger pigeon specimens. If you know of any, please contact the Project Passenger Pigeon website or the author in care of the publisher.

Almost all of the European countries have at least one. Other totals—Mexico: four birds in two museums; Australia: three birds in two museums; Japan: two birds; New Zealand: four birds in two museums; Guyana: one bird; and Sarawak, Malaysia: one bird. This last location stands out like a severe case of rhinophyma: How did a passenger pigeon get to Borneo? After a couple of letters to the Sarawak Museum failed to elicit any response, I contacted Frederick Sheldon of Louisiana State University, who conducts much of his research on Borneo. On his next visit, he met with the appropriate people in Sarawak and determined that, alas, no passenger pigeon was in the collection, nor any record of there ever having been one. But since Hahn received the information from the longtime director (now deceased), a passenger pigeon did undoubtedly once grace the collection. One can only imagine what happened to it.

George Lowery was a nationally prominent ornithologist who taught at Louisiana State University for many years. He was also a good friend of Schorger's, and they corresponded frequently. In May 1960, Lowery had just returned from New Zealand, where he was hosted most generously by a local ornithologist. They agreed to trade specimens, but the New Zealander badly wanted a passenger pigeon for his university's collection. Lowery sought Schorger's advice: "I now have a corner on the world's market of Whooping Crane eggs, thanks to the apparent infertility of good ole Crip and egg laying propensities of Josephine in the Audubon Park in New Orleans. Do you have any idea who might have an extra mounted specimen of a Passenger Pigeon to swap for a Whooping Crane egg?" Schorger responded by suggesting Lowery contact the Royal Ontario Museum, given its large collection of passenger pigeons: "Write Lee Snyder [of ROM] that nothing could promote the solidarity of the British Commonwealth more than the exchange of one for the egg of a Whooping Crane." That transaction was apparently never consummated, as ROM's only whooping crane egg is an early donation from James Fleming. But a trade was completed some years ago when a museum in Rhode Island gave up two passenger pigeons for a polar bear cub.

As one would expect, the amount of money it takes to acquire a passenger pigeon has increased dramatically over the years. In March of 1954 George Bachay from Edgerton, Wisconsin, wrote Schorger offering to sell one of his two "priceless" birds for $500. Schorger replied, "I . . . wish you luck in your attempt to get $500 a piece for the Passenger Pigeons. This figure is far beyond any that I have ever heard of. I bought a pair a few years ago for $75.00" (Schorger Papers).

No one keeps a closer eye on passenger pigeon sales than Garrie Landry, a botanist at the University of Louisiana–Lafayette. In 1982 he recalls a bird went for $500. Today, birds typically go for around $5,000, although on occasion specimens can still be had for closer to $3,000. The highest price Landry knows of that anyone has spent on a single bird in the United States is $12,000; this purchase occurred around 2003–04. As someone at Sotheby's explained, the price is determined by how badly a potential buyer wants the bird rather than any generally accepted value based on previous sales.

To this day a goodly number of birds remain in private hands, and more turn up regularly. Ideally, these birds would all wind up in educational or scientific institutions, through gifts, sales, or loans. Meanwhile, a registry should be established of these privately held birds. If anyone is interested, let me know via the Project Passenger Pigeon website or the publisher.

Bachay, George. March 29, 1954. Schorger Papers, University of Wisconsin Archives, Madison.

Baillie, J. L. "In Memoriam: Paul Hahn." *Auk* 82 (April 1965): 323–24.

Hahn, Paul. *Where Is That Vanished Bird?* Toronto: University of Toronto Press, 1963.

Extralimital

Archaeologists have discovered passenger pigeon bones in places well beyond their historical range. These include Dark Canyon Cave, New Mexico; Charlie

Lake Cave in northern British Columbia; Stansbury II site at the Great Salt Lake of Utah; and at least two sites in Southern California, including La Brea Tar Pits. Bird bones don't preserve well, so the presence of even a few in the far west suggests the possibility that the species once enjoyed a significantly wider range than that documented in the historical record. If that is true, the size of the total population might have been double or more the mind-boggling numbers that we do know about. Live birds were recorded in Wyoming, Idaho, Nevada, southern British Columbia, and southern Alberta.

The only documented instances of the bird in Mexico were the handful that showed up in the states of Veracruz and Puebla during the winter of 1872–73. Cuba hosted at least two passenger pigeons, a female shot in mangroves on the western part of the island and a male that turned up in a Havana market. Pigeons have been shot in France, England, Ireland, and Scotland, but Schorger doubted that any of the birds crossed the Atlantic on their own. Various people released passenger pigeons in all of those countries except possibly Ireland, but a species with the population and flight abilities of the passenger pigeon might well have made the trip on occasion, as have other North American birds.

To me, the most wayward of all passenger pigeons was the young male that appeared on Captain John Ross's vessel *Victory*, as she bucked in the heavy seas off Baffin Island on July 31, 1829. The bird arrived with a storm that roared out of the northeast. From that direction, the closest land was Greenland, as inhospitable to a passenger pigeon as most any place on earth. The waif might have been doomed, but it gave hope to the crew, as noted by Ross: "If the sailors called it a turtle dove, and hailed it as an auspicious omen, we were well pleased to encourage any of the nautical superstitions which served to keep up their spirits and furnish them with subjects of discussion."

Chandler, Robert. "A Second Record of Pleistocene Passenger Pigeon from California." *Condor* 84 (1982): 242.

Harington, C. R. "Quaternary Cave Faunas of Canada: A Review of the Vertebrate Remains." *Journal of Cave and Karst Studies* 73, no. 3 (2009): 162–80. doi:4311/jcks2009pa128.

Howard, Hildegarde. "Quaternary Avian Remains from Dark Canyon Cave, New Mexico." *Condor* 73, no. 2 (March–April 1971): 237–40.

Ross, John. *Narrative of a Second Voyage in Search of a North-West Passage and of a Residence in the Arctic Region During the Years 1829, 1830, 1831, 1832, 1833.* Brussels, 1835.

Serjeanston, Dale. *Birds.* New York: Cambridge University Press, 2009, 384–85.

Genetics

Despite the similarities in appearance between the mourning dove and the passenger pigeon, DNA analyses—based on material extracted from toe pads—indicate that the closest living relatives to the passenger pigeon are those in the New World genus *Patagionenas*, which includes the band-tailed pigeon of western North America. According to one study, it is likely that both species originated from one of the cuckoo doves of Asia (*Macropygia*). Under this scenario the birds crossed the Pacific into the Beringean Region (Alaska) during the Miocene, millions of years ago, although this would be one of the few instances known of North American land birds having Asian origins. A later paper suggested that these pigeons might have originated in the neotropics.

Fulton, Tara, et al. "Nuclear DNA from the Extinct Passenger Pigeon Confirms a Single Origin of New World Pigeons." *Annals of Anatomy*, 2011. doi:10.1016/j.aanat.2011.02.017.

Johnson, Kevin, et al. "The Flight of the Passenger Pigeon: Phylogenetics and Biogeographic History of an Extinct Species." *Molecular Phylogenetics and Evolution* 57 (2010): 455–56.

III. THE PASSENGER PIGEON IN HISTORY AND CULTURE

A. W. Schorger

Most of us have our heroes. For a long time now Arlie William "Bill" Schorger (1884–1972) has been one of mine. No one looms larger in the world of historical natural history. Schorger is the only person I know of who perused every newspaper ever published in a state in search of articles on wildlife (excluding waterfowl because there were too many stories on people hunting them). Even cutting off his searches at 1900, the exercise took fifteen years, produced 795 typed pages, and yielded a trove of information on the changing status of wildlife in Wisconsin that is unsurpassed for any other state. This remarkable database, augmented by other historical and scientific literature, enabled him to write a series of articles with such titles as "The Black Bear in Early Wisconsin," "The Prairie Chicken in Early Wisconsin," and "The Rattlesnake in Early Wisconsin." Out of this mass of material emerged two books as well, *The Passenger Pigeon: Its Natural History and Extinction* (1955) and *The Wild Turkey: Its History and Domestication* (1966). (His first book, published in 1926, is *The Chemistry of Cellulose and Wood.*) Of *The*

Passenger Pigeon, Schorger said that if the twenty-two hundred books and journal articles he read were added to the newspaper accounts, he consulted over ten thousand sources. Seven packed three-ring binders hold the notes from which he composed the manuscript.

Schorger was born in Republic, Ohio, in 1884 and received a master's degree in chemistry from Ohio State University. After a stay in Washington, D.C., he relocated to Madison, where he worked at the federal Forest Products Laboratory and pursued his doctorate in chemistry. Moving into the private sector, he started at the C. F. Burgess Laboratories in Madison and ascended the corporate ladder to become president of Burgess Cellulose Co., which had a factory in Freeport, Illinois. When he retired in 1950, he had to his credit thirty-four patents. But as one more testament to his lifelong interest in natural history and his affinity for data collection, he kept track of all the dead birds he encountered on his weekly drives between Madison and Freeport. He made the trip 693 times and recorded 4,939 individuals of 64 species.

During Schorger's short tenure at the Forest Products Laboratory, he met and befriended another young scientist, Aldo Leopold. The two of them formed a group with several others who also enjoyed canoeing, fishing, hunting, and other outdoor activities. Leopold was one of the great thinkers and writers of the twentieth century. His *A Sand County Almanac* is an amazing amalgam of science and philosophy expressed in some of the most beautiful and powerful English prose you will have the pleasure to read.

It is impossible to say how Leopold's career would have evolved had he left Wisconsin, but Schorger and other friends succeeded in ensuring that such a move never happened. In 1933, the University of Wisconsin hired Leopold as the nation's first professor of game management, and he joined the Agricultural Economics Department. Research by Stan Temple and Curt Meine (who authored the definitive biography of Leopold) revealed that coincident with the hiring, Schorger and other group members made substantial donations to the university. It is thought that these funds went toward Leopold's salary. The same nucleus of patrons stepped up when Leopold was offered a high-level federal position in Washington, D.C. Due to their lobbying and financial generosity, the school created a whole department for him: Leopold was the sole faculty member of the country's first Department of Wildlife Management.

After his retirement from business, Schorger joined the department in 1951 as professor of wildlife management. He evidently team-taught but one course, and he earned a token amount rather than an actual salary. Four years later he became an emeritus, but remained a presence at the depart-

A. W. Schorger. Courtesy of Stan Temple and the University of Wisconsin Department of Forest and Wildlife Ecology; photo by Tim Wallace

ment until the year before his death in 1972. He left an estate in excess of $6 million, at the time one of the largest in Dane County history. Fifty thousand dollars went to the department to help add books for the library, and a like amount established a scholarship for the study of Italian art, a subject of great interest to his wife. Marie McCabe explained, with a dollop of humor, that at Schorger's death, colleagues heard about the endowment and praised the generosity, but when they later learned how much the estate was, the accolades turned to grumbling.

Schorger married Margaret Davison, the daughter of a small-town banker, in 1912. Margaret was physically large and possessed a strong personality: two different people who knew her described her as "formidable." One next-door neighbor suggested that Bill was quiet because Margaret was so loud. She was apparently the only reviewer of his written work whose suggestions he accepted. In gratitude and with humor, he dedicated the passenger pigeon book to Margaret, "whose patience surmounted extinction." She died in 1961, a loss he struggled with for the remaining decade of his life. They had two sons, William and John. William served on the anthropology faculty at the University of Michigan for many years, and John taught writing and language at the Minnesota Metropolitan State College.

Two stories are worth relating, for one pertains to Schorger's tenacity as a natural historian and the other as a businessman. Wisconsin's only extant specimen of a cougar was an individual killed near Appleton in 1857. For many decades its mounted carcass graced the biology department of Lawrence University. But the faded cat was eventually deaccessioned into the trash, from which it was rescued by a tavern owner who thought it would make a nice addition to his establishment. It would undoubtedly have stayed there until the next remodeling had not the story been recorded in an old newspaper that Schorger examined as part of his ongoing research. He drove to the bar and for $50 liberated the scarred and odoriferous hostage. It now resides in the beer- and smoke-free confines of the University of Wisconsin's zoology collection. In honor of Schorger's efforts, another Badger, Hartley Jackson, christened the animal a new subspecies, *Felis concolor schorgeri*, a designation no longer recognized.

Toward the end of World War II, the U.S. military found itself low on sleeping bags, due to shortages of the feathers and kapok that were used as fillers. A request went out for proposals for substitutes. Schorger believed he had found one: cattail seeds. He rented a large warehouse at the corner of University and Whitney Way in Madison and hired students to collect masses of cattail heads. Apparently the building was literally brimming with the plants and could hold no more. But before Schorger had concocted an adequate technique for processing, other sources of stuffing became available and the bedding crisis passed. For Schorger, though, a new challenge arose: the dried cattail heads had discharged their seeds, and every inch of the building was covered in fluff. It was described as a sleeping bag with a tin exterior.

Schorger was meticulous in his research and his life. Data ruled and details mattered. He would spend hours caring for his lawn, removing dandelions by hand. When someone had the temerity to report a rare bird at the Kumlien Club, Schorger would pepper him with questions, the key one often being whether a specimen was procured. One story goes that upon learning of a brown pelican (an oceanic bird that occurs only rarely inland) on Lake Mendota, he raced out to collect it, only to encounter a group of birders who were not happy at the specimen slung over his shoulder. Once he decided that the perfect dessert was blueberry pie à la mode, he had no need for further experimentation, and blueberry pie finished nearly every meal.

Robert McCabe and Joseph Hickey, close friends and colleagues of Schorger's, both comment on how serious he seemed: no photos show him smiling, he was gruff, had no patience for ineptitude, and was difficult to know. (At least he smiled and laughed in person—unlike Leopold who has been described as almost grim in his countenance.) McCabe and Hickey found these impressions lamentable-because they knew the real Schorger possessed a first-rate sense of humor. *The Passenger Pigeon* and *The Wild Turkey* were each criticized for their lack of personal analysis and overabundance of citations and "historical and biological statistics." Professor McCabe made the point that the body of the passenger pigeon book ends with this sentence: "A photograph of a nest with an egg occurs in Craig." Still, even in a work written in the Joe Friday style ("Just the facts, ma'am"), there are gems of jocularity that reflect the author's viewpoint. Perhaps my favorite is on page 85. After quoting the Reverend David Zeisberger to the effect that a foot of dung accumulated at a roost in a single night, Schorger commented, "It is to be noted in the history of this pigeon, data involving the highest figures are given by men of the cloth, a trait not inconsistent with a belief in the miraculous."

In reading through the boxes of Schorger's correspondence housed at the University of Wisconsin archives, I was struck not only by how genuinely funny he was but also how considerate to the many strangers who wrote him. A high school senior from Northfield, Ohio, asked Schorger if he could provide her with information on "cyclic reactions in growth and development of plants and animals in the United States," the subject of her term paper. After confessing he did not really know what she meant, he provided a potentially helpful reference. A letter from a father in Richland, Washington, asked about the wisdom of his son's majoring in wildlife management. Schorger acknowledged that the field offered limited opportunities, and that "the decision should be based entirely on his personal desires." (He might have suggested that the prudent course was the one he followed: first, amass a fortune, then pursue the nonlucrative calling.) In July 1968, a correspondent from Livonia, Michigan, wrote a rambling, and in places incomprehensible, letter linking nesting passenger pigeons in Guatemala that were driven north to Michigan by a volcanic eruption in the 1890s and Persied meteors with 1960s grizzly maulings in Montana and mass murder in Saskatchewan. Maybe by then Schorger had mellowed a bit, for he promptly responded, "This will acknowledge receipt of your interesting letter . . . As to their nesting in Guatemala, there is no record of the passenger pigeon south of central Mexico."

Ornithologist Ralph Palmer, then at the New York State Museum, sent a short note on January 16, 1961, expressing his interest in visiting Madison: "How is Lake Mendota on or about April 14? Should I bring my bathing suit, or can such be rented locally?" Schorger replied, "Lake Mendota is always open by April 14 but contains ice floes that make the swimming lively. I might add that bathing suits are prohibited, the reason being that no one should feel inhibited."

(In October 2009, I interviewed several Schorger acquaintances due to the kind assistance of Stan Temple. This piece is, in part, based on those interviews with Emily Early, Marie McCabe, Phil Miles, Gene Roark, and Stan Temple. Ms. Early has since passed away.)

Hickey, Joseph. "In Memoriam: Arlie William Schorger." *Auk* 90 (July 1973): 664–71.

McCabe, Robert A. "A. W. Schorger: Naturalist and Writer." *Passenger Pigeon* 55 (1993): 299–309.

Schorger, A. W., Papers. University of Wisconsin Archives, Madison. To Palmer, January 16, 1961; to Dzuro (student), November 11, 1963; to Livonia, July 31, 1968.

Economics

Economists have used the history of the passenger pigeon to demonstrate why the market system does not prevent extinction. These demonstrations go down two paths. First, with scarcity the price ought to increase. Now that could either depress demand or fuel it, with more hunters trying to cash in on the higher prices and consumers wanting the prestige of buying rare things. (People are willing to pay a premium for acquisition of the higher status that they think comes with spending lots of money: this is known as the Veblen effect.) But in the case of the passenger pigeon, the market was unable to distinguish passenger pigeons from other game, or even domestic poultry, so as the availability of pigeons declined, the overall supply of cheap meat did not. Therefore, passenger pigeon prices neither rose nor did demand fall. Further, because of the few large massings of the birds in the early 1880s, the true rarity of the species was masked, so it was easy for most to assume that there were still plenty.

The second economic principle embodied in the destruction of the passenger pigeon relates to the impoverishment of the commons, a term that could mean a resource open to all or one that is open to a specific group. But given the wandering nature of the species, only national governments could effect limits on the exploitation of the passenger pigeon as a resource, and those laws came too late for this bird. Cornell University economist Jon Conrad wrote a paper entitled "Open Access and Extinction of the Passenger Pigeon in North America," in which he uses models and formulas to explore the economic factors that led to the bird's demise. Conrad's work was summarized at my request by Jeffrey Sundberg, professor of economics at Lake Forest College: "A thriving market for passenger pigeons, technology that allowed for low search costs, low shipping costs, and high remuneration (in comparison with other jobs the hunters could perform), and a low opportunity cost of wage for farmers combined to make extinction a logical outcome, given that no property owner limited access to the resources."

Conrad, Jon. "Open Access and Extinction of the Passenger Pigeon in North America." *Natural Resources Modeling*, 2005.

McDaniel, Carl, and John Gowdy. "Markets and Biodiversity Loss: Some Case Studies and Policy Considerations." *International Journal of Social Economics* 25, no. 10 (1998): 1454–65.

Perelman, Michael. *The Natural Instability of Markets*. New York: St. Martin's Press, 1999, 53–55.

Tober, James. *Who Owns the Wildlife?* Westport, CT: Greenwood Press, 1981.

Eugenics

It takes a far more imaginative mind than mine to connect the extinction of the passenger pigeon with eugenics, but the oddest document in Professor Schorger's papers is the undated Pamphlet No. 58 published by the "successor to Eugenics Society of Northern California": "We had no wildlife conservation a century ago. Now we strenuously try to save our last whooping crane, our ivory-billed woodpeckers, our roseate spoonbills. Fine. But how about talented humans? With excessive birth control, our irreplaceable leadership types are going the way of the dodo, moa, the great auk, yes . . . even yesteryear's passenger pigeon."

Memorials to the Passenger Pigeon

I am aware of four memorials to the passenger pigeon: (1) In 1947, the Boy Scouts of America dedicated a memorial in the Pigeon Hills of New Hanover, Pennsylvania. It was destroyed by vandals in 1961 and was rededicated on September 12, 1982, at Codorus State Park, where it overlooks Lake Marburg; (2) also in 1947, the Wisconsin Society for Ornithology erected a passenger pigeon monument at Wyalusing State Park in Bagley, Wisconsin (at a ceremony the year before, Aldo Leopold read his essay "On a Monument to the Pigeon," one of the most poignant ever written about extinction); (3) to mark the Petoskey, Michigan, nesting of 1878, a memorial was installed at the Michigan State Fish Hatchery at Oden, Michigan, in 1957. The 1878 flocks had nested over a three-county area, with a large concentration at Crooked

Lake, near the hatchery; and (4) the Cincinnati Zoo has preserved the aviary that housed Martha, even spending thousands of dollars to move it a short distance when zoo renovations threatened its existence. Distinguished artist John Ruthven aided the zoo in their preservation efforts by helping raise funds through the sale of prints made of a special passenger pigeon painting he created. He also donated an antique shotgun to the exhibit, never dreaming that someone would break in, steal the weapon, saw off most of the barrel, load it with modern shells, and attempt a robbery. The guy was caught, and at his trial Ruthven had to testify that it was indeed his gun and explained how it wound up as the weapon. The gun was returned and John filled the barrel with lead so it could never be used as a weapon again, except possibly as a club. That is the gun on display today.

Music

1. Opera

The Dresden (Germany) Music Festival commissioned Deborah Artman to write the libretto, and the San Francisco–based Bang on a Can, comprising the three composers Michael Gordon, David Lang, and Julia Wolfe, to write the music, for a unique exploration titled *Lost Objects*. It was first performed in May 2001. Artman explains that she and the three composers looked to Jewish texts and tradition for inspiration and direction: "The Talmud attempts to define—how lost things bind all people together, how we build our life around things that have been lost and forgotten, or lost and not forgotten." The pieces range from consideration of the mundane, "I Lost a Sock," to extinction, "Passenger Pigeon":

Passenger Pigeon
was once
one of the most
numerous birds
on earth
Thousands of pigeons
carrying messages of
sport
carrying messages of loss
carrying messages of life
No matter how long it is
gone
No matter how far it

has flown
the bird
will always come home.

(The text for Julia Wolfe's "Passenger Pigeon," libretto by Deborah Artman, is reprinted by kind permission of Red Poppy Music.)

2. Popular Music

At least three songs focus on the last passenger pigeon. The best known of the songs is the highly sentimental "Martha (Last of the Passenger Pigeons)" by the late giant of bluegrass music John Herald. This touching work, as printed lyrics, is best appreciated by skimming the words without pausing at the mention of "dozers" in the 1870s and the bestowing on Martha of certain intellectual and emotional characteristics that have not yet been documented in birds:

Oh high above the trees and the reeds like rainbows
they landed soft as moonglow
in greens and reds they fluttered past the windows
ah but nobody cared or saw

till the hungry came in crowds
with their guns and dozers
and soon the peace was over
God what were they thinking of?

Oh on and on til dreams come true
you know a piece of us all goes with you.

Oh the birds went down
they fell and they faded to the dozens
Til in a Cincinnati Zoo was the last one

Yes all that remained was the last
with a name of Martha
Very proud, very sad, but very wise.

Oh as the lines filed by there were few who cared
or could be bothered

how could anyone have treated you harder
and it was all for a dollar or more.

Oh on and on til dreams come true
you know a piece of us all goes with you

Oh and surrounded there by some of whom wept around her
in a corner of the cage they found her
she went as soft as she came so shy til the last song
oh the passenger pigeon was gone.

(Reprinted with permission.)

A humorous take on the last passenger pigeon was penned by the Canadian naturalist/musician David Archibald (1994, Rogues' Hollow Music). Many liberties were taken with the facts, most glaringly turning the last bird into a male. But the goofy tone creates an appealing result in "The Passenger Pigeon's Lament":

I was never that beautiful to look at
Still, there were times I was at my best
The ladies would all turn their beaks to see me
When I puffed out my red and manly chest.

They'd keep a staring right above my shoulder
To the favorite part of my anatomy
Where the feathers all were purple, green, and golden
A finer pigeon neck you will never see.

CHORUS

But where oh where have my buddies gone
Where are my family and friends
We used to block the sun
But now I am the only one,
I'm the last of the passenger pigeons.

Sure, we got complaints our nests were always messy
That branches could not stand the heavy load

Now, I wish that I'd inquired what the price was
For contravention of the nesting code.

For they hunted us like dogs, well, more like pigeons
With nets and sticks and guns, they were so rude
Its distracting hearing all those shouts of "Timber"
When you are trying to get your loved one in the mood.

CHORUS

My relatives enjoyed a balanced diet
Be it beechnuts, winter green, or raspberries
For a true gourmet's delight, you ought to try it
With a farmer's freshly planted field of peas.

But soon the "family dinner" changed its meaning
Now, what's a lonely pigeon going to do
When his cousins are all smoked and dried or roasted.
And Uncle Walter's (that's Walter Pigeon) always in a stew.

CHORUS

I'm the last passenger pigeon
'Bout to cross the waters Stygian
And I'm hopin' that religion sees me through
'Cause there's no one left to care now
For I leave behind no heirs now
I'm alone and in despair now
Yes it is true
I'm just living out my days inside this zoo.

(Reprinted with permission.)

The Handsome Family in their album *Twilight* (2001) give an accurate rendition of passenger pigeon history in a poignant metaphor for lost love, "Passenger Pigeons":

Ever since you moved out
I have been living in the park

I'd rather talk to the wind
Than an empty apartment.
And I wish I could forget
How a billion birds flew in
My hollow dying heart
The first time I touched your arm.

Once there were a billion passenger pigeons
So many flew by, they darkened the sky
But they were clubbed and shot
Netted, gassed, and burned
Until there was nothing left
But vines of empty nests
I can't believe how easily
A billion birds can disappear

The park is empty now
It's so cold out
And all the paddle boats
Are covered up with snow.

Once again it is dark
The electric lights snap on
But I'm still sitting here
Drinking frozen beer
And throwing potato chips
Into the white snow drifts
Just in case a bird decides
To fly through hinter night

I can't believe how easily
A billion birds can disappear
Oh, I can't believe how easily
A billion birds can disappear.

(Permission granted.)

* * *

Scouse the Mouse is a delightful children's album featuring Ringo Starr as the main performer, with Donald Pleasance as the producer and author of the lyrics and Roger Brown writer of the music (released in Great Britain in 1977 by Polydor Records). Among the vast literature of the English language, the song "The Passenger Pigeon" is undoubtedly unique in featuring verse that rhymes *platypus* with *Ectopistes migratorius.*

Novels

At least three novels devoted to the passenger pigeon appeared in the twentieth century, two by well-known and highly acclaimed authors. The first and most unusual is published in 1938 by MacKinlay Kantor (1904–77), *The Noise of Their Wings.* (The title is a quote from Audubon.) E. D. Starke, sickly as a child, is sent by his parents to stay one summer with an uncle and aunt who operate a farm in Michigan. Just after midnight one morning, Starke's uncle rouses the boy from his slumber and forces him to participate in a raid on a passenger pigeon roost. The event leaves an indelible scar on his psyche. When the young man grows to become head of a giant food-canning operation, he devotes substantial portions of his fortune to conservation and scientific efforts. He never loses hope that some passenger pigeons still survive all these years later. With an independence of mind bolstered by great personal wealth, he announces that he will give $100,000 to anyone who can provide him with a living pair of passenger pigeons. Amazingly, a legitimate claim emerges from the Gulf coast of southern Florida. Most of the action takes place when Starke assembles his estranged daughter, a longtime friend who is an ornithologist, and others to receive the birds. A dispute with one of the claimants for the reward results in arson and the death of the passenger pigeons. Kantor includes a bibliography that contains Forbush, Mershon, and French, and what he says about the birds is mostly factual. *Time* (October 31, 1938) said the story "is teasing and ingenious rather than effective" and that the author spent most of his "vitality . . . devising a modern plot." But I respect Kantor for creating a novel that incorporates the passenger pigeon into a plot that goes beyond straight natural history; he reaches out to a potentially larger audience to share the pigeon's history with those who might not otherwise know anything about it. Kantor wrote thirty books, many of which dealt with the Civil War, including the Pulitzer Prize–winning *Andersonville.*

The other two books are nature novels, of the type where the author tells the life history of a species through the story of one individual. *The Last Passenger* by James Ralph Johnson was published in 1956. According to the book's jacket, he was born in 1922 and spent his career in the U.S. Marines. At the time of the novel, he held the rank of major and was teaching in an ROTC program at the University of Louisville. He had written two previous nature books for children. This slim book tells the life of Blue, a male passenger pigeon born in a huge gathering on the Ohio River. Blue wanders north to another large nesting near Petoskey, Michigan, but excessive disturbance by hunters forces the birds to flee northward, where they wind up nesting near Hudson Bay. (Too many nestings in a year for my taste.) In his meanderings across pigeon range he encounters a host of wildlife including Carolina parakeets, ivory-billed woodpeckers, and Labrador ducks; survives the impact of a diving peregrine falcon; finds a mate; spends time as a stool pigeon; and flies to freedom when shooters at a trap meet fail to bring him down. The manuscript was sent to Schorger, who gave his approval, and the *New Yorker* called it "a rare, unpretentious, and, in its way, singularly appealing work."

Allan Eckert's *Silent Sky* was published in 1965. The last passenger pigeon of this novel is also a male, one with a narrow splotch of white on his right wing. He was born in Michigan and would end his days at the hand of a child in southern Ohio; in death his eyes would be replaced with buttons. I think Eckert is a more accomplished writer than Johnson, and Eckert's bird, spared a name, has fewer harrowing escapes. Although Eckert takes some liberties with the details, he works into his story more of the historical literature, including a passage on Roney's account of Petoskey. More people were probably introduced to the story of the passenger pigeon through *The Silent Sky* than any other single source.

Eckert holds a rare place in American letters: he wrote mostly about natural history and Midwestern history, two realms usually far removed from popular culture, yet enjoyed a broad, national audience. Of his many books, seven were nominated for Pulitzer Prizes in literature. He also authored 225 episodes of the long-running nature program *Wild Kingdom*. His highly regarded play *Tecumseh!*—based on the life of the Shawnee prophet—has been performed in Chillicothe, Ohio, every summer since it premiered in 1973. I wrote him in the fall of 2011 to tell him about Project Passenger Pigeon and learned that he had died the previous July. His wife said he would have been interested in our effort, and I regret not having contacted him earlier.

From 2010 through 2013, at least four novels were published with major plot elements involving passenger pigeons: *Quick Fall of Light* by Sherrida Woodley (Spokane, WA: Gray Dog Press, 2010) is a superb science fiction novel dealing with a number of environmental themes; *Post* by Hilary Masters (Kansas City, MO: BkMk Press, 2011) is a funny satire connecting the disappearance of species with the impoverishment of culture (among other things); *Chase the Wild Pigeons* by John Gschwend Jr. (self-published, 2011), is an adventure novel about two boys in the Civil War trying to find their way back home; and *One Came Home* by Amy Timberlake (New York: Alfred Knopf, 2013) is an admirable novel for young adults that combines mystery, adventure, and humor against the backdrop of the huge pigeon nesting of 1871.

Place Names

The website www.placenames.com makes it easy to find all the locations in the United States that have *pigeon* as part of their names. Given the low regard that most people hold for feral rock pigeons, the word *pigeon* in most geographic entities within the historical range of the passenger pigeon likely refers to this species (one exception is Pigeon House Corner, a populated place in Ann Arundel County, Maryland). Pigeon Forge, Tennessee, is probably the best known, but there are Pigeon Creeks throughout the eastern third of the country. Here is a sample of states and the number of towns, schools, churches, hills, creeks, and other places within their respective borders having appellations that include *pigeon*: CT—4; TN—26; KY—44; WV—38; GA—16; PA—32; WI—26; OH—20; MI—20; MO—17; IN—16; NY—13; MA—12; IL—12; LA—9; and IA—5. Margaret Mitchell lists a number of pigeon bays, lakes, islands, rivers, points, and even a rapids in Ontario and Manitoba.

White Pigeon, Michigan, and White Pigeon, Illinois, are actually named for a Potawatomi chief who saved a white settlement by alerting the inhabitants to an imminent attack. But his name does refer to the bird in question. Huntingburg, Indiana, memorializes a nearby passenger pigeon roost that attracted hunters for decades. The name of Mimico, a Toronto suburb, derives from the nineteenth-century Mississauga word, *omiimiikaa*, denoting a place where wild pigeons gather. Ontario is the home of two other places based on Indian names: Omemee in Victoria County, and Omemea, an island in the Parry Sound area of Georgian Bay. (Variations of a similar word meaning passenger pigeon appear as *o-me-me-wog* in the language of the Potawatomi and *omimi* among the Cree and Chippewa.) There is one high-profile place named after the passenger pigeon, in Quebec: Île aux Tourtes (Passenger Pigeon Island) which is connected to Montreal by a high-traffic bridge (Pont de l'Île-aux-tourtes).

Poetry

Amos Butler ended his article on the passenger pigeon in his *Birds of Indiana* (1897) with these words: "Their passing away must fill the soul of every one, into whose life their migrations have come as an experience, with profound regret. I introduce the lines of a careful observer, a faithful interpreter of nature, my friend, Hon. B. S. Parker. His "Hoosier Bards" are the feathered songsters of our beloved State, and therein he has preserved his recollections of the Passenger Pigeon:

And windy tumults shake the ground,
And trees break down with feathered store,
And many swiftly pulsing wings
Are spread at once in sudden fright,
Till every fleeting minute brings
The noise of some delirious flight,
And all the air is dark with swarms
Of pigeons in their quest for food,
While autumn leaves in eddying storms
Are beaten by the feathered flood.

Written in the 1880s or 1890s is the poem "Wilda Dauwa" (Wild Dove) authored by the Reverand Eli Keller, who lived in Northampton County, Pennsylvania, and knew the birds well. It is written in Pennsylvania German and was translated by Alan Keyser at my request (via Rudolf Keller). It is presented here in English for the first time:

In olden times there were passenger pigeons
We saw them fly in spring time
In small flocks and charming large ones
What a great pleasure it was!

The boys in the fields a plowing
Would stop with their tired teams
And in full voice call "pigeons."
You heavenly beautiful creatures.

And they flew away over mountains high,
Still higher over deep valleys,
So, we, with joy, just let them fly:
Thought to ourselves, "You may decide."

We hear guns cracking here and there
Rusty iron long-time loaded.

The shooting is worth nothing—just noise,
Damaging the lazy shooters themselves.

Still finally the pigeons tired from flying
Set down with water rushing
In cool shade consider themselves lucky
There they call their "Eht" and listen.

How beautiful they are sitting in long rows,
And in high green trees and branches,
With little gray caps and tidy gray coats
With red and white vests!

They are now entirely gone these wild pigeons,
Eternally never to come back!
What yet remains of these beautiful blessings?
The spirit lays its treasures down.

One of the greatest contemporary authors of children's literature is Paul Fleischman, a Newbery Medal winner and the 2012 U.S. nominee for the international Hans Christian Andersen Awards. In several of his poetry collections, he writes in two voices, including "The Passenger Pigeon." This poem appeared in his collection entitled *I Am Phoenix: Poems for Two Voices* (published by HarperCollins, 1985, and used here with permission from the publisher):

<table>
<tr><td>We were counted not in</td><td></td></tr>
<tr><td></td><td>thousands</td></tr>
<tr><td>nor</td><td></td></tr>
<tr><td></td><td>millions</td></tr>
<tr><td>but in</td><td></td></tr>
<tr><td>billions.</td><td>billions</td></tr>
<tr><td></td><td>We were numerous as the</td></tr>
<tr><td>stars</td><td>stars</td></tr>
<tr><td></td><td>in the heavens</td></tr>
<tr><td>As grains of</td><td></td></tr>
<tr><td>sand</td><td>sand</td></tr>
<tr><td>at the sea</td><td></td></tr>
<tr><td></td><td>As the</td></tr>
</table>

buffalo	buffalo
	on the plains.
When we burst into flight	
	we so filled the sky
That the	
sun	sun
was darkened	
	and
day	day
	became dusk.
Humblers of the sun	Humblers of the sun
we were!	we were!
The world	
inconceivable	inconceivable
	without us.
Yet its 1914.	
And here I am	
alone	alone
	caged in the Cincinnati Zoo,
the last	
	of the passenger pigeons.

Arts Etobicoke is a collective of artists active in West Toronto that has presented a wide range of innovative offerings to the community and region. In October 2010, they unveiled the Art Alley Mural Project, which helps celebrate the Universal Declaration of Human Rights. Article 13 addresses the freedom of movement. The project uses both a painting and a poem by Toronto's poet laureate Dionne Brand called "Article 13." Although the passenger pigeon is not central to the poem, Brand uses the birds as both a symbol of place (*mimico,* the name of a nearby suburb, means "where wild pigeons gather") and the freedom to migrate. But although the words refer to people, it is a pretty fair description of the bird itself, "tributaries of migrants / inalienable nomads," making uncounted sojourns.

The novelist William Burroughs, a keystone of the beat movement of the 1950s, produced a unique collection of writings, much of it imbued with sardonic humor and unleavened bleakness. His poem "Thanksgiving Day, Nov. 28, 1986" is a litany of America's flaws and begins: "Thanks for the

Drawing of Martha and accompanying text by unknown artist that graced a wall of the Stockholm, Sweden, subway. Many of the train stations there feature permanent and temporary art exhibits.
Photo by Anna-Karen Granberg

Ski Country Limited Edition Whiskey Decanter with porcelain passenger pigeons, issued as a special collectable in 1983. It is based on the original artwork of Donald Leo Malick. Courtesy of Garrie Landry

wild turkeys and passenger pigeons destined to be shit out of thoroughly wholesome American guts."

Paintings and Sculptures

The earliest European drawing of the passenger pigeon is one with a forked tail that appears in Louis Nicolas's *Codex canadensis*, seventy-nine pages of text and pictures that depict the people and natural history of the eastern half of the United States and Canada. The work was prepared about 1700.

Mark Catesby is the painter of the first colored illustration of the species. There have been innumerable drawings since, manifesting a vast range of talent, styles, and settings, including trading cards; dishes; faux gravestones; advertising; puzzles; a subway wall in Stockholm, Sweden; a mural in the Dennison, Ohio post office; several postage stamps issued by such nations as Mozambique, Cuba, and Tanzania; and as tattoos on at least three people. I have already discussed Lewis Cross, whose passenger pigeon drawings may be unique in that he knew the birds from life and created his dramatized images specifically to remind people of their former multitudes. Some well-known contemporary painters who have incorporated images of the species in their work include Norman Rockwell, John Ruthven, Walton Ford, and Hunt Slonem. Kate Garchinksy of Philadelphia and Tim Hough of Toronto are two young artists who have created beautiful images of passenger pigeons.

A male passenger pigeon appears on one of the plates in the state dinner service that was commissioned for President and Mrs. Rutherford B. Hayes in 1879. Mrs. Hayes selected Theodore Davis, who worked for *Harper's Weekly*, as the artist to oversee the production of the dishes. He created images that represented native animals and plants, including the pigeon.

Sculptors have re-created images of the birds as well. Todd McGrain, for example, has made one that has been placed at the Grange Insurance Audubon Center, on the banks of the Scioto River south of Columbus, Ohio. Another artist is Rosalind Ford, who lives in Newfoundland and Labrador and creates textile sculptures of flocking birds including passenger pigeons.

Photographs

One of the most perplexing aspects of the passenger pigeon slaughter was the apparent absence of all photographic documentation: no photos of a guy with one dead pigeon, none of wagons filled with dead birds, none of barrels of pigeons being lined up along a railroad track or a Great Lakes dock, none of a bird or two hanging from the stalls of a game market, none of coops with live birds to be used for food or shooting contests, or of contestants in urban trap meets showing off their dead birds.

Schorger never published such a picture nor are there any in his papers. Scholars such as Garrie Landry, Stan Casto, and Stan Temple have searched for years without success. There is a photo in Mershon of a game market with dead birds, but none is clearly a passenger pigeon. A photo on the Internet shows a huge pile of bison skulls that is sometimes purported to be of pigeons. Garrie had a false alarm once when he read on eBay that a seller was offering a photo of dead passenger pigeons. He contacted the person and was allowed to examine the photo, but, alas, when he looked at the picture with a microscope, he could see that the alleged passenger pigeons had webbed feet, making them most likely teal or other species of small duck. Supposedly, too, a photographer was present at Kilbourn City (Wisconsin Dells) at the time of the 1871 nesting, but the story goes he was so appalled by the slaughter he deliberately refrained from recording it. This seems to be almost certainly apocryphal, but for whatever reason, the pictures he did take do not include passenger pigeons. And unfortunately the half-tone process that allowed newspapers to print photos had not yet been developed when the pigeons were still in the wild.

Finding a picture of newly slain wild passenger pigeons became an important goal of mine as I was convinced that such pictures must have been taken. I contacted hundreds of people and institutions throughout Canada and the United States in my search, but to no avail. A number of interested librarians and archivists joined the effort and continued their own investigations. But none of us struck pay dirt.

Then, on October 9, 2012, I received this remarkable e-mail from Destry Hoffard: "I saw in the recent issue of *On Target!* [newsletter] that you were looking for photographs that showed Passenger Pigeons in a trap shooting or hunting sense. I've collected vintage photos for years and only have two that might fill the bill." He sent me copies, and sure enough they are indeed passenger pigeons, as the reader can see (photo insert). To my knowledge both are unlike any photos previously published. (Although, quite amazingly, on January 17, 2013, my colleague Susan Wegner spotted two versions of the same stereopticon card on two websites so obviously multiple copies are still in existence. That different companies featured the same shot suggests it sold fairly well.)

Destry recalls obtaining the stereopticon card about a decade ago at a postcard show in Toledo. There are more details associated with his acquisition of the tintype. He was in northern Michigan, on his way to Iron Mountain, when he stopped for lunch and noticed a small antiques shop nearby. Destry asked the proprietor if he had any old hunting photos. After rummaging about for a while, the owner emerged with a box of tintypes, out of which he selected one: "The light in the shop was bad and I didn't really look too closely. He wanted the princely sum of $3 and it seemed like some kind of sporting scene so I just rolled the dice and bought it. It didn't really dawn on me what it might show till I got it home a few days later and began to study it with a glass."

The stereopticon card is entitled "Small Wild Game of the Alleghenies" and depicts a string of gray squirrels, below which are a number of ruffed grouse and three passenger pigeons. It was produced as part of the "Stereoscopic Gems of American and Foreign Scenery" series by the Universal Photo Art Company (1880–1910), headquartered in Philadelphia. The use of the image represents the common practice of reissuing photos taken earlier. In this instance, the photographer was likely R. A. Bonine of Altoona, who took many photos on commission from the Pennsylvania Railroad Company during the 1870s.

The tintype (photo insert), probably taken in the 1860s but possibly as late as 1880, has a number of fascinating elements that are difficult to identify with absolute certainty. An important point is that this was a staged grouping (as evidenced by the painted background) and not a shot of people actually in the field catching passenger pigeons. In addition, the pigeon paraphernalia would have varied greatly, as it was not manufactured but rather made by individual hunters. I have discussed this image with a number of people, particularly Garrie Landry and art historian Susan Wegner of Bowdoin College, and the latter undertook a detailed analysis of the picture. The long pole held by the man on the left is the rod that would be attached to a fulcrum stake allowing it to be moved up and down teeter-totter like. The disk at the top is the stool where the decoy pigeon was perched. The short cross stake below the man's hand was driven into the ground under the stool and provided the lowest point that end of the teeter-totter could reach; depending on the nature of the ground, this would prevent possible damage to the stool and the pigeon. A rope from the pole enabled the pigeoner to move it up and down. The long box with the holes—probably of wood with a canvas top—on which the rightmost pigeon preens is likely the traveling cote that held the stoolies and the fliers. (The basket might have served the same purpose.) Finally, the bag with the lumpy contents probably held the nets.

The only picture that comes close to that of the two men and their stool pigeons is the one that appeared in an article long known to passenger pigeon researchers, but the significance of the photo was not realized until February 2012 during a conversation between Garrie Landry and me. In his paper "The Last of the Wild Pigeon in Bucks County," Colonel Henry Paxson included a photo of longtime pigeon trapper Albert Cooper of Solebury Township, Bucks County, Pennsylvania, posed with what appear to be three live passenger pigeons. These birds are described in the caption as "blind decoys," referring to the practice of temporarily blinding stool pigeons by sewing their eyelids shut during their use as decoys. The photo was taken "about 1870."

There are other photos of live birds, but over 90 percent of them are of birds in Professor Whitman's Chicago flock. The remainder, minus the two already discussed, are from the Cincinnati flock, and most, if not all, of those are of Martha. I have never seen a photo of Whittaker's Milwaukee birds.

RADIO, TELEVISION, MOVIES, AND THEATER

References to passenger pigeons have appeared from time to time in radio, television, movies, and theater. On April 27, 1948, the *Fibber McGee and Molly* radio show aired the episode "The Passenger Pigeon Trap." McGee thought he found a living passenger pigeon despite the generally held view that the species was "stinct." The bird in question, a rock pigeon, was perched on a bus and hence was a passenger.

In the inaugural episode of *Star Trek*, entitled "Man Trap" and aired on

September 18, 1966, a visit to a planet virtually devoid of life elicits mention of the extinct bird. Running from 1982 to 1988, the PBS children's show *3-2-1 Contact* had a regular feature involving the Bloodhound Gang, a group of kids who solved mysteries. "The Case of the Dead Man's Pigeon" centered on a contested will that is proved fraudulent because one provision leaves the estate to a society dedicated to the conservation of passenger pigeons; since the bird is extinct, there can be no such society and the will is a forgery. The script was written by acclaimed children's writer and Newbery Medal–winning Sid Fleischman, whose son Paul carried on the family tradition of writing superb children's literature, including the passenger pigeon poem printed above.

In Jim Jarmusch's movie *Ghost Dog: The Way of the Samurai* (1999), a character refers to a particular bird as a "carrier pigeon" but is corrected (incorrectly as it turns out) by the owner, who yells, "Passenger pigeon! Passenger pigeon! They've been extinct since 1914!" Terrence Malick's *New World* (2005) shows unmistakable faux Carolina parakeets, and one scene has a cloud of birds in the background that some have claimed are passenger pigeons, but a viewing of the movie indicates they could be any of the birds that flocked in huge numbers during that early period of U.S. history.

Passenger Pigeons, a feature film dealing with a coal-mining town in eastern Kentucky, was released in 2010. Writer-director Martha Stephens answered my query regarding the title: "I named the film *Passenger Pigeons* because . . . [it] represents the death of a culture that was once flourishing [but] is a much different animal than what it once was. Sadly, poverty still reigns but the work ethic and a sense of self-pride has seemed to disappear much like the Passenger Pigeon. I guess to sum it up, my movie touches upon the extinction of a place and the extinction of a person" (e-mail, April 7, 2011).

In the early winter of 1999, the Kraine Theater in New York City presented the play *American Passenger* written by Theron Albis and directed by Stephan Golux. Ostensibly about a family of New York gangsters, it explores the consequences of an inability to change when circumstances demand it. Mr. Golux kindly elaborated on the content of the play in response to my question: "*American Passenger* is a multi-layered fugue on the dangers of the failure to adapt. The title of the play refers directly to the American Passenger Pigeon, mentioned explicitly in the script and referenced implicitly in the production as the metaphor for a species that cannot evolve to meet rapidly shifting threats to life" (e-mail, April 5, 2011).

Satire

An interesting and slim volume of ornithological satire was produced by Melissa Weinstein and Jack Illingworth in 2003, entitled *The Writings of Noah Job Jamuudsen on the Passenger Pigeon* and purportedly published by the National Jamuudsen Society. The authors, having met in Montreal, discovered the text of this previously unknown ornithologist, the letters of whose name are coincidentally the same as those of America's most famous bird student, "hidden inside a cask of St-Ambrose Oatmeal Stout." Born in Sweden in the early 1800s, Jamuudsen came to the United States sometime before the summer of 1867, for that is when he first appears in "the archives of the Rhode Island Brotherhood of Ornithologists [as having been] involved in a number of disturbances." It is the goal of Weinstein and Illingworth to place the subject of their biography on his "rightful throne: that of the Priest-King of all ornithologists." Jamuudsen's account of the passenger pigeon is indeed unique in the extensive literature on this species.

Urban Legend

A totally fabricated story about the final days of the wild passenger pigeon has permeated the Internet so thoroughly that it has been repeated as fact by serious authors in numerous articles and books (including one by Stephen Jay Gould). Like the most persistent of urban legends, it sounds plausible, is a narrative, and contains a moral. The claim is made that the last of the pigeons, all 250,000 of them, attempted to nest near Bowling Green, Ohio, in 1896: "In a final orgy of slaughter over 200,000 pigeons were killed, 40,000 mutilated, 100,000 chicks destroyed." Only 5,000 managed to survive the wrath of the hunters. The 245,000 corpses were processed and packed in barrels for rail transport east. Unfortunately, however, the train derailed and all the dead birds spoiled and had to be discarded.

Unless one is steeped in the passenger pigeon record, it would be easy to swallow this story, but red flags are everywhere. First, as I document in the second-to-last chapter, there were nowhere near a quarter of a million birds in 1896. By then, all that still survived in the wild likely numbered no more than maybe a hundred or so. Second, not only don't the numbers add up, but the purported breakdown of avian casualties is both far too precise and too high a percentage of the total to be believable. Third, and of great importance, such an event could not have escaped the knowledge of every historian and ornithologist who has ever written about the species. It would have made big news at the time. Finally, it is just too neat a story that every one of the dead birds was wasted. (The same article says a flight of over two billion passenger pigeons flew over Cincinnati in 1870; again, no such thing has ever been reported anywhere else.)

This fable is a classic type of folklore known as the cautionary tale, which usually has three elements: a taboo or bad thing to do (killing all the birds); violation of the taboo; and a bad consequence (birds spoiled and discarded). The amazing thing to me, though, is that people are still making stuff up about a bird that has been extinct for nearly a century. I have attempted to trace this fable to its origins, and I think—although I am not sure—it first appeared in an anonymous article called "The End of the Wild: An Essay on the Importance of Biodiversity," which supposedly ran in the now defunct *Borealis* magazine. The language is elegant and written in the first person; the author claims to have shared podiums with the likes of Richard Leakey. But modern authors who refuse to use their names and offer no explanation are suspect from the get-go. You can find the tale at http://raysweb.net/specialplaces/pages/endofwild.html.

Acknowledgments

I started my research on this book in the fall of 2009. Besides combing the literature with the help of several librarians, I solicited information on Publore and other specialist listservs including the birding listservs of close to twenty states. This netted a host of valuable materials and, more important perhaps, contacts with interested people who have proved to be strong allies throughout this ongoing effort. I spent nine days in Madison, Wisconsin, going through the boxes of Schorger's personal papers in the University of Wisconsin's Archives and the binders filled with his passenger pigeon notes that are kept at the Department of Forest and Wildlife Ecology. There were also trips to numerous other places through former passenger pigeon range. Throughout this I was aided by a host of people, a number of whom have become good friends. The following is a partial list of those who provided information, reviewed chapters, provided lodging, or provided other assistance that made completion of this book possible (some names appear in the text):

Bob Adams, Glenn Adelson, David Aftandilian, Renee and David Baade, Paul Baicich, Kyle Bagnall, Mark Barrow, Eleanor Bartell, Fred Baumgarten, Craig Benkman, David Blockstein, Charlie Bombaci, Ken Brock, Simon Bronner, Neely Bruce, Alan Bruner, Michael Bryson, Blair Campbell, Angelo Capparella, Stan Casto, Elisabeth Condon, Bill Cook, Mary Cummings, Bob Currin, Stan De Orsey, Julia Di Liberti, Jim Ducey, Tim Earle, Josh Engel, Nancy Faller, Ralph Finch, Raymond Fogelson, Beryl Gabel, Paul Gardner, Nancy Gift, Nancy Glick, Bob Glotzhober, Ben Goluboff, Stephen Gordon, Don Gorney, Terri Gorney, Michel Gosselin, Robin and Travis Greenberg, Nadia Gronkowski, Greg Hanson, Stan Hedeen, Lynn Hepler, Destry Hoffard, David Horn, Richard Horwitz, Mary Hufford, Connie Ingham,

Julia Innes, Paul James, Carolyn Jaskula, Jeanette Jaskula, Kevin Johnson, Jacqueline Johnson, Kenn and Kim Kaufman, Tom Kastle, John Kay, Rudolf Keller, Tom Kent, Cynthia Sue Kerchmar, Alan Keyser, Kindler family, Gretchen Knapp (reviewed entire manuscript), Deborah Lahey, Garrie (reviewed entire manuscript) and Lynn Landry, Cindy Laug, Ellen Lawlor, Leo Lefevre, Bob Levin, John Leonard, Wendy Lilly, Karen Lippy, John Low, Damon Lowe, Peter Ludwig, Karen Lund, Dan Marsh, Jerry Martin, Terrance Martin, Bob Maul, Carol McCardell, Todd McGrain, Curt Meine, Ed Meyer, Janet Millenson, Steve Mirick, David Mrazek, Greg Nobles, Ben Novak, Mark Peck, Justin Peter, Wayne Peterson, Pamela Rasmussen, Steve Rogers, Kayo Roy, Sarah Rupert, Robert Russell, John Ruthven, Rita Rutledge, Sarah and Steve Sass, Linda Scarth, Jonathon Schlesinger, Jennifer Schmidt, Theresia Schwinghammer, David Scofield, Geoffrey Sea, Skip Shand, Jerri Sierocki, Andy Sigler, Mathew Sivils, Arthur Smith, Michael Solomon, Kathleen Soler, David Stanley, Tom Steel, Nancy Steiber, Wendy Strothman, Steve Sullivan, Jeff Sundberg, Stephanie Szakal, Stan Temple, Jeremiah Trimble, Sophia Twichell, Tim Wallace, Jason Weckstein, Susan Wegner, Daniel Weinman, Bill Whan (reviewed entire manuscript), Phil Willink, Jon Wuepper, Owen Youngman, and Kristof Zyskowski.

A few of the names on this list have appeared in the acknowledgments of my other books as well. I cannot express the full level of my appreciation for these long-term friends whose love and other support have helped make all of this possible.

Notes

CHAPTER 1: LIFE OF THE WANDERER

1. Craig (1911) 408; New York minister in Wicks 108.
2. Craig (1911) 410; a life on the wing in French 80; John James Audubon 320.
3. Hudson Bay in Mitchell 92; Mississippi in Lincecum 194–95; Maryland in E. Grant 28.
4. "Pigeon hosts" in *Forest and Stream* (1913) 792; Narragansett in Schorger (1955) 254.
5. Mitchell 84.
6. A. Wilson 108.
7. "Flight was very rapid" in Mitchell 84; Heriot in Wright (1911) 350.
8. Clait Braun, e-mail, July 21, 2011.
9. Fleming in Mitchell 169.
10. King v–vi.
11. Ibid., 121–22.
12. Schorger (1955) 201–02. The three scientists referred to are John Leonard, University of Illinois (Chicago), Ken Brock, and Stan Temple, University of Wisconsin–Madison. The first two looked at the question at my request.
13. Ibid., 204.
14. Kalm 61; Bishop 54.
15. Lincecum 194–95.
16. Butler in Leonard 165–67. See Kirtland 68.
17. Forty-two genera in Schorger (1955) 36–45; Mitchell 101–02.
18. Williams 145.
19. Ibid., 143. See Lalonde and Roitberg 1303 and Sork et al. 528–41.
20. Schorger (1955) 126–27; Ellsworth and McComb 1554.
21. Thoreau in Cruikshank 104; and "bag of marbles" in Roberts 585.
22. Pokagon in Mershon 59.
23. Benkman, e-mail to author.
24. Bertram 76.

25. "Blue wave" in Schaff 107; "rolling cylinder" in Wheaton 442; fecundity of forest in Scherer 32.
26. Scott County in *Viroqua Censor* (WI) and *Indiana Farmer*; Tennessee in Wright (1911) 442.
27. Cook in Mershon 164–66.
28. *Cumberland Daily News*. Schorger did not think it credible that the pigeons would just sit there getting squashed as their comrades in ever-increasing quantities rested on their backs. He thought it more likely that each bird perched on a branch that bowed downward to make it appear the birds were actually on top of each other. But the illusion, if such it was, presented itself often.
29. Webber 305–08.
30. Ibid.
31. Casto 11.
32. Ibid., 11–12.
33. *Indianapolis Star*. Also in G. Wilson 16:442.
34. Tennessee in Wright (1911) 442; Black River in Hall 56–58.
35. Hall 56–58; chemical release in T. M. Harris 179–80; air was so impregnated in Revoil 8.
36. Gonterman 1–50. This is a virtually forgotten work, overlooked by Schorger and others. From the passion Gonterman expresses in his introduction—"the extermination of the passenger pigeon . . . is a disgrace to civilization"—it is easy to surmise that the story of the bird had invaded his youthful consciousness as it has so many who know it. And being from Kentucky, he had probably read the accounts of his state by Wilson and Audubon a century before and become intrigued by what had happened to the roosting places. Through questionnaires and interviews, he solicited from the old-timers firsthand information on the local status of the bird, including the location and size of the roosts.
37. Schorger (1955) 87.
38. Seton 523. See Atkinson 7.
39. Coale 254–55.
40. Mershon 50–51.
41. Craig (1911) 420–21. See Whitman (1919) 120.
42. Seventy to a hundred twigs in Mershon 205; structures often persisted in Schaff 107.
43. Pike in Wright (1910) 436; "military precision" in French 56; "avenues . . . one mile" in French 12–13.
44. Josselyn in Wright (1910) 436; New Hampshire in Wright (1911) 358; Sparta in *Fond du Lac Commonwealth*.
45. Lincecum 194–95.
46. Giraud 184–85.
47. French 49.
48. *Fond du Lac Commonwealth*.
49. Deane (1896) 236. Deane attributed the tilting to incubation during cold spells when the bird's wing would cover the eggs for warmth. But pigeon experts David Blockstein and Garrie Landry both doubt that the pigeons would have incubated in that way, as no other bird does. The passage is based, however, on Deane's

firsthand account published in the *Auk*, so I don't doubt he described what he saw. Landry provided the plausible explanation adopted here.

50. *Detroit Post and Tribune.*
51. Nutritious milk in Hegde 238; milk-laden squab in Dixon.
52. Failure to dispense in French 58; Martin (1879) 385.
53. "Like drunken men" in *Whitewater Register* (WI); they hiked their way in French 30–31; human disturbance in Godwin 176–78.
54. Schorger (1955) 125.
55. 1976 in Greenberg (2002) 402; 1740 in Kalm 57; Schoolcraft 381.
56. Marsden 146–47.
57. J. J. Audubon 35; Welsh 165–66. Audubon refers to foxes in Kentucky, which at that time period would likely have been gray rather than red.
58. Goshawks are the most brazen in Bent 133; J. J. Audubon 242.
59. Trautman 209–10.
60. Wrong kind of hit in Bertram 70.
61. Scott 9–10.
62. Kelly 339–42. See Ostfeld et al. and McShea.
63. The American-chestnut was a mast producer that comprised 25 percent of the trees growing in the Appalachian Mountains. Chestnut blight (*Endothia parasitica*), an airborne fungus from Asia first imported into the United States around 1905, infected trees at a rate of fifty miles a year. (This had no impact at all on passenger pigeons, for by 1905 the species was likely gone from the wild.) Over the next few decades, an estimated thirty billion trees died, practically the entire population. Mast-dependent organisms must have been devastated during the period between the disappearance of the chestnuts and their replacement by other trees. From an ecological perspective the loss was even more significant in that the chestnut varied much less in the quantity of its annual mast than did the oaks and hickories that replaced it. At least one study of chestnuts, based on estimates, found that over ten years chestnuts did not experience a single mast failure, and that the nuts produced during low-production years were just less than half of what appeared during high years (Diamond et al. 196–201). The forests were more stable when American chestnuts still lived. Unlike the chestnut blight, gypsy moths were deliberately brought, to Medford, Massachusetts, in 1868 to create disease-resistant hybrids with native silk moths. Some, however, escaped their confines. They found the New England woodlands hospitable enough to become a serious problem by 1889, when the first outbreak was discovered. They have since spread into the Midwest, where they are now firmly established. Gypsy moths are cyclical and can at their peak denude oaks of their foliage, eventually killing them. A major moth predator is the white-footed mouse, though *The Yearbook of Agriculture 1949* took comfort in the demonstration that the aerial application of DDT effectively controlled the moth.
64. Webb 367–75.
65. Ellsworth and McComb 1548–58.
66. Ibid.
67. Ibid., 1553.
68. Noss 234.

69. Ibid., 235–36.
70. "One of the most disastrous" in Kriska and Young 3; ideal-size food in Raithel 21–23.
71. Life cycle of tick in Ostfeld et al. 326; Blockstein (1998) 1831. See Jones et al. 1023–32.
72. Komar and Spielman 164.
73. This discussion on lice is based on Clayton and Price; Dunn; and Friederici.

CHAPTER 2: MY BLOOD SHALL BE YOUR BLOOD: INDIANS AND PASSENGER PIGEONS

1. Interview with Paul Gardner, Midwest Regional Director, Archaeological Conservancy.
2. Krech 183.
3. Interview with Terrance Martin, Illinois State Museum.
4. Parmalee (1959) 62–63.
5. Guilday 1.
6. Guilday and Parmalee 163–73.
7. Orlandini 73–75.
8. Jackson 186.
9. Krech 36–37; S. Nelson 8–16; Neumann 389–410.
10. Parmalee (1958) 174.
11. Mitchell 17; Wright (1910) 429.
12. Mooney 280; Mitchell 18.
13. Fradkin 415–16.
14. Hager 92–103.
15. Kalm 64.
16. Schorger (1955) 140.
17. Dodge 343.
18. Krech 24.
19. Lawson 50–51.
20. Radin 112.
21. Atkinson in Blanchard 159.
22. Bunnell 186.
23. Jackson 177.
24. G. Harris 450.
25. The remainder of the chapter is based on the remarkable paper by Fenton and Deardorff 289–315.

CHAPTER 3: A LEGACY OF AWE

1. Cook 17–18.
2. Wright (1911) 361; O'Callaghan 45.
3. Schorger (1938) 473.
4. Ibid., 471, 475.
5. Wright (1910) 431, 434.

6. Wright (1910) (Stork) 435; Watson 410.
7. Mathew 1–3.
8. See Powell.
9. A. Wilson 399.
10. Ibid., 400.
11. J. J. Audubon 322.
12. Ibid., 320–21.
13. M. Audubon 200–03. She presents both what Audubon wrote of Wilson and the exact quote of what Wilson wrote about Audubon.
14. Elsa Guerdrum Allen wrote, "Wilson's greater exactness, his patient method and his lucid and honest descriptions mark him unquestionably as the better ornithologist." As for the feud, Schorger (in Scott 12) commented, "Audubon did not relish another star in the ornithological firmament and his treatment of Wilson does not rebound to his credit."
15. Wallace Craig, who studied the only flock subjected to scientific scrutiny, found Audubon's account so full of errors he spent the last page of his paper criticizing it in detail.
16. French 162–63.
17. Mershon 49.
18. Nevins 181–82.
19. Raper.
20. *Cleveland Plain Dealer.*
21. Newspapers in Schapper 102; "very abundant" in Nelson 120; 1881 in Butler (1898) 763.
22. *South Shore Country Club Magazine* 34.
23. Eenigenburg 12–14.
24. McKenney 352–53.
25. F. 149–50.
26. Wharton 1–2.
27. Dodson clippings.
28. Stratton-Porter 196–98.
29. Franklin xxxviii–xxiv.
30. Robinson 568.
31. Cooper chapter 22.
32. Winter online 1–7.
33. Clarke 266–76.
34. Upton 1–4.
35. Ibid., 21, 50.
36. Ibid., 105.
37. Hewitt 82–85.
38. Upton 231–32.
39. Ibid., 231.
40. Ward.
41. Lake Shore Museum Center archives.
42. Ward. Also see Holland, Michigan, *Evening Sentinel.*

43. Ward; *Muskegon Chronicle* 1937.
44. *Muskegon Chronicle* 1951.

CHAPTER 4: PIGEONS AS PROVISIONS TO PIGEONS AS PRODUCTS

1. Laudonniere 114; Heriot in Schorger 5; Champlain in Biggar 332.
2. Powers 105–11.
3. Mitchell 21, 106.
4. Wright (1911) 356.
5. 1759: Wright (1911) 435; A. Wilson 401.
6. Bourne 563–64.
7. Boston and Granby: Judd 351–52; Schorger (1955) 131; Benwell 72–75.
8. French 177.
9. DeVoe 175–76.
10. Ibid., 172–73; Byrd: Wright (1911) 432.
11. M'Neill in G. Wilson 14:570; E. Wilson (1934) 164.
12. Mitchell 21.
13. Madison: Hamel 18; Chicago: *Home Guide.*
14. Nessmuk 106; 1770s in Wright (1911) 350; Althouse in Mitchell 107.
15. Roasted pigeons in Faux 22; Belknap 137–38; Illinois in French 184.
16. L. Thomas 105–110.
17. Ibid., 93–95, 98–101.
18. Soap in French 206; Coudersport in Thompson 14.
19. Saint-Jérôme in Schorger (1955) 132; Ontario in Mitchell 108; McKnight in Schorger (1955) 132.
20. Brickell 186; Native healer in Larocque 49.
21. A. Wilson 399; British ornithologist in Schorger (1955) 52; West Virginia in Brown 176.
22. Jesuit and La Hontan in Mitchell 16.
23. *Prairie Farmer* 83.
24. Michigan in *Cass County Republican* 3; Pennsylvania in Rupp 131; Eden, Wisconsin, in *Fond du Lac Reporter*; Iowa in Bond 525.
25. Goss in Roberts 586.
26. Kalm 63; Van Campen in Armstrong 10.
27. Massicotte 77.
28. Wisconsin in F. E. Jones; Hine 327; Minnesota in Swanson 116–17.
29. Schorger (1955) 53.
30. Jasper County in *Galveston Daily News* (1875) 2; Ibid., (1881) 3; Leon County in Casto 13.
31. Wright (1910) 431.
32. Quebec in Mitchell 62; Hussey 5.
33. Cook in Mershon 167–68.
34. Texas in Casto 14; Ontario in Mitchell 109–12.
35. Mitchell 109–12.
36. MacKay 262.

37. R.
38. Harpel 205.
39. During 1850s in Answers.com 2–4; Wisconsin in Price 19; as early as 1842 in Schorger (1955) 144–45.
40. French 177; *Grant County Herald.*
41. 1842 in French 98; 1880: ibid., 103.
42. Mershon 124–25.
43. Roney 346; Allen brothers in Mershon 125.
44. McKinley 407.
45. Pennsylvania in Mershon 126; Phillips: ibid., 109.
46. French 213–15.
47. Original receipt in possession of Milwaukee Public Museum. They provided me with a copy.
48. Forbush (1913) 99.
49. Armstrong 4; Bennett in Traverse 1411.
50. Price 35–36.
51. Merritt 27–31, 109.
52. St. Paul in Swanson 63; Merritt 111.
53. Merritt 113.
54. Ibid., 184–85
55. BMR 395–96.
56. Coale (1922a) 255.
57. C. L. Mann 45–47.
58. Forbush (1913) 99–103.

CHAPTER 5: MEANS OF DESTRUCTION

1. Schorger (1955) 167–68.
2. Wright (1911) 436.
3. Tennessee in Wright (1911) 443; Texas in Casto.
4. St. Lawrence River in Wright (1910) 430; St. Paul in Swanson 133; Orillia, Ontario, in Mitchell 129.
5. Stone 488.
6. Trautman 271.
7. Mitchell 120–21.
8. Hussey 5.
9. Twain 114.
10. Grant 42–43.
11. Mitchell 120.
12. Schorger (1955) 196–97. According to the *New England Weekly Journal* of April 8, 1740, a mill near Philadelphia "took fire and burned to the ground" likely due to "the Wadding of Guns fired at Wild Pidgeons."
13. St. Paul in Swanson; Cabot in Brewster (1906) 176.
14. Swivels in Randolph 95; Mather in Schorger (1938) 473; 1662 in Wright (1910) 430; 99 birds in Mitchell 122.
15. Mitchell 119.

16. Kalm 66; New York in *Milwaukee Sentinel*; 1860 in French 48–49.
17. Wright (1911) 350.
18. Webber 305–08.
19. Sage 69–70.
20. Mitchell 123.
21. "Snap Shot" 194.
22. *New England Weekly Journal*, April 8, 1740, 2.
23. Schorger interview with Victor Blasezyk, May 31, 1936. All of the shooting incidents recorded here from Wisconsin appear in newspaper clippings collected by Schorger and placed in his Passenger Pigeon Notebooks (two volumes of "Pigeons: Wisconsin Newspapers"), located at the Department of Forest and Wildlife Ecology, University of Wisconsin, Madison.
24. Ray.
25. Randolph 95–96.
26. 1870s in Trautman 270; lured into pens in Brewster (1889) 289; stuffed pigeons in Wright (1911) 352.
27. Ontario examples in Mitchell 128–29; Mather in Schorger (1938) 473.
28. Armstrong 5.
29. Early in the season in Mershon 108; pigeon baskets in *Antique*, cover and Baillie notes.
30. Yarnell in Deane (1931) 264–65; Ontario in Mitchell 124; linen in Rupp 133.
31. Typical rig in French 195. See Snyder 10–13, Lincoln, Mitchell 124–27, and Rupp 133–35.
32. French 227.
33. White River in Mershon 109; Benzie County in Maynard 241–42.
34. F.E.S. 50.
35. Sibley 414.
36. Finer grain in Rupp 134; Wisconsin in *Milwaukee Journal* (1929); angleworms in Barrows 242.
37. Garber 28.
38. Fancy model in W. W. Thompson 15. See Lincoln, Paxson 376–77, and Scherer 40.
39. Found suitable birds in Paxson 378; "John (X)" 299; $5 to $10 in Rupp 133; began his exercise in W. W. Thompson 13.
40. John (X) 299; Scherer 40.
41. W. W. Thompson 16–17.
42. Rupp 132.
43. Beekmantown in R.; Osborne in Mershon 127; Michigan: ibid., 109; Dr. Voorheis in Barrows 245.
44. French 153, 210.
45. Scherer 42; French 177.
46. French 102.

CHAPTER 6: PROFILES IN KILLING

1. *St. Joseph Traveler*.
2. Competition for fun in Leffingwell 133; ecological context in Price 33. The shooting matches were not without risk, however. On one occasion contestant Hiram

Neiswinter missed his pigeon and struck rival Robert Parker instead. "The top of Parker's head was blown off and his brains was [*sic*] scattered all around." (*The Carbon Advocate* [Leighton, PA], August 19, 1882, 51.)

3. Field shooting in Leffingwell 135; one set of skills in Swanson 258; "character, coolness" in Price 29.
4. E. Thomas 369.
5. Leffingwell 139.
6. E. Thomas 369.
7. Leffingwell 42, 143.
8. Ibid., 136.
9. Rosenthal in Schorger (1955) 160. See Swanson 259, Mitchell 115, and Steele 220.
10. Ontario in Mitchell 117; April-through-September in E. Thomas 372; *Forest and Stream* in De L. 233.
11. Casto 16.
12. *Galveston Daily News* (1884) 2.
13. Swanson 258–60.
14. Kennicott Club in *Chicago Tribune* (August 17, 1872) 6; "grand tournament": ibid., (September 30, 1877) 7; Peoria in Schorger (1955) 163.
15. E. Thomas 371–72.
16. Greenberg 358–59.
17. Bogardus 300.
18. Ibid., 301–02.
19. Ibid., 302–03.
20. *Chicago Tribune* (September 27, 1872) 6
21. Ibid., (May 2, 1880) 3.
22. Bogardus 302–03.
23. Schorger (1955) 163.
24. Price 33.
25. Steele 3.
26. *Dictionary of Unitarian and Universalist Biography.*
27. Steele 143, 150.
28. Ibid., 220.
29. Ibid.
30. Ibid., 233.
31. Ibid.
32. *New York Times* (June 21, 1881) 5; *New York Times* (June 22, 1881) 2.
33. *New York Times* (June 22, 1881) 2.
34. E. Thomas 372.
35. Czech 26.
36. Joyce 10–15.
37. Wilson, "Kin-ne-quay," 1.
38. Ibid., 2.
39. Ibid., 3–4.
40. The remainder of this section on Wilson is based on Wilson (1934) and Wilson (1935). This latter note includes the discussion of Partie.

41. "G. D. Smith Succumbs."
42. This section is based on George D. Smith's unpublished memoir, 1–6. See also Rumer.
43. John French's book, in two intriguing sentences that include no further elaboration, also touched on the social aspects of pigeon hunting, but the glimpses provided are decidedly dark and foul. The first is from his pen and appears on page 20: "There are camp-fire stories galore of the carnivals of the slaughter and the orgies of the feasts, when the day's work was finished, that are better buried in the oblivion of silence." And the second is from his publisher and editor, Henry Shoemaker, which appears on page 173: "Added to the horrors of squab hunting and killing were orgies of drunkenness that made the scenes in the nesting grounds too hideous to recount." One can only imagine.
44. *Chatfield Democrat* (May 13, 1865), (June 3, 1865).
45. Swanson 144–45.

CHAPTER 7: THE TEMPEST WAS SPENT: THE LAST GREAT NESTINGS

1. Mitchell 109–11. This discussion was substantially aided by the input of the following Ontario experts: Glenn Coady, Nicholas Escott, Michel Gosselin, George K. Peck, and Mark Peck.
2. Fox 102–03.
3. Scherer 38–39.
4. Ibid.
5. Ibid., 41.
6. Ibid.
7. Schorger (1937) 1.
8. Ibid., 4–6.
9. Ibid., 19–20.
10. *Kilbourn City Mirror* (April 22, 1871).
11. Six hundred in Mershon 117–18; one hundred thousand in *Fond du Lac Commonwealth*; sixteen in Schorger (1937) 17–18; icehouse in Mershon 113.
12. H. Kelly.
13. *Kilbourn City Mirror* (May 13, 1871).
14. This paragraph and the next three are from *Fond du Lac Commonwealth*.
15. Schorger (1937) 12–13.
16. Ibid., 13–14.
17. Roberts 583.
18. Swanson 70.
19. Howland 1976.
20. Hartwick and Tuller 80–81.
21. Ibid., 81.
22. "Tom Tramp" 149.
23. Mershon 106–09; Hartwick and Tuller 81.
24. *Michigan Tradesman*.
25. Souter: ibid; one estimate in Hartwick and Tuller 81.

26. Martin (1914) 478–81.
27. Hartwick and Tuller 81.
28. This and the next two paragraphs from Fenton and Deardorff 314–15.
29. Forty miles in length in Roney 345; “trollops” in Sharkey 6.
30. *Charlevoix Sentinel* (March 12 and 25, 1878).
31. Timber operators in *Northern Tribune* (March 9, 1878) Bemis in *Charlevoix Sentinel* (April 19, 1878).
32. *Emmet County Democrat* (March 29, 1878).
33. Petoskey: ibid. (April 5 and 12, 1878); Cheboygan in *Northern Tribune* (April 13, 1878).
34. Charles.
35. Bennet in Sharkey 13; Old Joe in Hedrick 54–55.
36. Peterson 48–49.
37. *Laws of Michigan* 149–50.
38. Roney 345–49.
39. Ibid.
40. Ibid.
41. Ibid.
42. Ibid.
43. Ibid.
44. Ibid.
45. Ibid.
46. Ibid.
47. Numerous observers in Lawrence and Henkel 25; dismissal of all charges in *Emmet County Democrat* (May 3 1878), *Charlevoix Sentinel* (April 26, 1878), and Turner 401–2.
48. *Emmet County Democrat* (April 19, 1878); Sharkey 8.
49. Sharkey 9.
50. Roney 346; Martin (1879) 385–86; another pigeon merchant in Mershon 93.
51. *Potter Journal* (April 15, May 13, and June 10, 1880).
52. *Detroit Post and Tribune* 1.
53. Ibid.
54. Ibid.
55. Ibid.
56. Barrows 215.
57. Mershon 56.
58. Ibid., 56–57.
59. Littlefield 154–59.
60. *Chicago Field* 314–15. The Atoka account is based on this article.
61. Files of *Potter Journal*; Thompson 6.
62. Scott 14.
63. Dixon.
64. Adams County in *Baraboo Republic*; hired two hundred in *Daily Data*.
65. Scott 15–16; *Daily Data*.
66. *Milwaukee Evening Wisconsin*; Schorger (1955) 217.

CHAPTER 8: FLIGHTS TO THE FINISH

1. Casto 10.
2. McKinley 410. Daniel McKinley, through his painstaking study of Missouri's newspapers and other historical documents, is responsible for much of what is known about the bird in that state.
3. Ibid, 411.
4. W. W. Thompson 6.
5. Oconto River and Racine in Schorger (1955) 218; Oviatt in Scherer 30.
6. French 87.
7. Ibid.
8. Ibid., 87–88.
9. Ibid., 59–60.
10. 1869 in Blockstein and Tordoff; Griscom 212–16. Accounts of passenger pigeons abandoning young before they were fledged go back a long ways. It seems likely that this was in response to human disturbance, although it was not recognized as such. But starting in the late 1860s, the propensity to leave a nesting site early in the face of disturbance seems to have increased, perhaps in part because it was more discernible among smaller concentrations of birds.
11. Blossburg in French 61; Bailey in Roberts 584; Missouri in McKinley 411.
12. Wisconsin in Schorger (1955) 219; Pope in Casto 6; W. Cook 248.
13. Brewster 285–86. A small nesting of several hundred birds was reported from West Virginia in 1889.
14. Missouri in McKinley 412; Arkansas in Hough (1892) 138.
15. Norfolk in GH 79; Clinton County in Todd 271; winter 1892–93 in Forbush (1913) 76; New York City in Fleming (1907) 236–37.
16. Virginia in Stanstead 403; Iowa in Anderson 239; New Jersey in W. Stone 154; Tennessee in Schorger (1955) 292.
17. Clark 44.
18. Ibid.; Woodruff 88.
19. Indiana in Butler (1898) 763–64; Wisconsin in Schorger (1955) 221; Massachusetts and North Carolina in Forbush (1927) 75.
20. Schorger (1955) 286. If someone had been fortunate enough to have seen a passenger pigeon in the late 1890s, or even worse the early 1900s, he would have had to choose between killing a bird whose very existence was imperiled or have his sighting rejected. Of course the death of a single individual is not likely to cause extinction, although if all the "single individuals" are aggregated, the issue might become a bit blurred with some species, but the rarity of the target ought to weigh on the consciousness of a potential collector. And one could argue that the life of an endangered bird is more important, anyway, than getting credit for seeing something exceedingly scarce. It is also true, though, that just as the loss of an individual won't determine the fate of a species, especially one such as the passenger pigeon, science doesn't gain much either in determining whether the pigeon became extinct in the wild in 1900 or 1902 or 1906. But this exercise in historical sleuthing is compelling nonetheless, for it is worthwhile trying to make sure that the end of a great and tragic story is as accurate as possible.

21. Miller and Griscom 130.
22. Mandeville in Forbush (1927) 75; Wisconsin in Schorger (1955) 220; Michigan and Illinois in Deane (1896a).
23. Nebraska in Deane (1896a) 81 and Bruner 84; North Western in Johnson; Clinton in Eaton 385; Ontario in Fleming (1907) 236–37; West Virginia in Buckelew; Jones in Roberts.
24. Shannon in Butler (1898) 764; Texas in Simmons 86; Ontario in Fleming (1903) 66; New Jersey in Chapman 341; Maine in Palmer 299; Wisconsin in Hollister 341; Iowa in Widmann 85 and Anderson 239; Pennsylvania in Paxson 372 and Schorger (1955) 291.
25. Louisiana in McIlhenny 546; Missouri and Pokagon in Deane (1897) 316–17.
26. Neither of these records have previously been published, and I appreciate the assistance of Steve Sullivan at the Peggy Notebaert Nature Museum and Jeremiah Trimble of the Museum of Comparative Zoology.
27. Mitchell 137; Jeremiah Trimble, personal communication.
28. *Osprey* 12; November in Beckner 55–56.
29. Hough (1899) 88; Schorger (1955) 208.
30. Fleming 66; Moody 81.
31. J. Wood 208; N. Wood 225.
32. Eaton 386; Atkinson 8.
33. Butler (1899) 150 and (1902) 98–99.
34. Dr. McGrannon in Todd 271; Little Rock in Litzke 24; Babcock in Hough (1910).
35. Schorger (1955) 286.
36. Henninger 82; Geoffrey Sea, personal communication.
37. Offered to donate in Cokinos 232; *Ohio Conservation Bulletin* 17.
38. Cokinos 244.
39. Forbush (1927) 77; Townsend 379–80. Schorger rejected six post-1900 specimen-based reports that may well have been identified correctly, even if some of them are lacking important details. This judgment is based not only on the people who claimed to have seen the specimens but the stature and reliability of their contemporaries who assessed and accepted those claims. The first two of these involve a St. Louis game dealer who told Otto Widmann, Missouri's leading ornithologist at the time, that he received twelve dozen birds from Arkansas in 1902 and another in a shipment of ducks from Black River, Missouri, in 1906. Widmann thought it unlikely that the dealer would have erred in his identification. Pennsylvania state ornithologist Harvey Surface told the legislature in 1904 that he received a bird the previous year that had been shot out of a flock of seventy-five or eighty. Maine is home for a 1904 record that is based on the observation of a newly killed bird in a taxidermy shop. Forbush and Ora Knight (Maine's leading ornithologist of his day) found the record credible, as have Ralph Palmer and Peter Vickery. "A Swede" reportedly shot a passenger pigeon at North Bridgeport, Fairfield County, Connecticut, in August 1906. It wound up in the collection of George Hamlin, which is now housed at the Natural History Museum of Los Angeles County. Fleming sank the record by pointing out that Hamlin never mentioned the bird when he contributed to a 1913 book on Connecticut birds, an assertion that dismisses the possibility that Hamlin acquired it later. And finally,

in 1915, octogenarian J. L. Howard sent a bird to Cornell University along with a letter providing a detailed account of his having shot it in 1909. The mount bore a date of 1898, which could have referred to a previous tenant, but Fleming argued Howard was too old to remember whether an incident occurred six years ago or seventeen. In addition to these birds, there is a specimen in the Yale University collection from Bay City, Michigan, that is dated January 24, 1906. But after an examination of the evidence, Kristof Zyskowski (Yale's bird collection manager), Jon Wuepper (editor of *Michigan Birds* and *Natural History*), and I all agree that the 1906 date clearly refers to when the donor received the bird and not when it was killed. And last, a bird supposedly shot in Chicago in 1901 could just as easily have been killed in 1891 (Greenberg (2002) 507).

40. Menard County Illinois History.
41. Purdue 51.
42. Butler (1902) 98–99.
43. Butler (1912) 64. I thank Dr. Stan Hedeen for alerting me to the 1912 paper and Bill Whan for the 1902 paper.
44. Schorger (1955) 223.
45. Merriam in Cutright 152–53; Burroughs in Brinkley 686.
46. Mershon 185, 179.
47. French 172.
48. Hodge (1911) 49–50.
49. *American Field* (1910) 124–25.
50. Ibid., 125.
51. Ibid.
52. Nests also found in Hodge (1911) 51; Harrington in Anonymous.
53. Hodge (1912) 169–74.
54. Ibid., 174.
55. Ibid., 175.

CHAPTER 9: MARTHA AND HER KIN: THE CAPTIVE FLOCKS

1. French 180.
2. *Milwaukee Journal* (September 18, 1935), *Milwaukee Journal* (June 14, 1898).
3. HM. 539.
4. Deane (1896) 235–37.
5. Ibid., 236.
6. Ibid.
7. Ibid., 237.
8. Morse 271.
9. Pauly 145–46.
10. Ibid., 162.
11. Deane (1908) 181–183; Whittaker 30.
12. Deane (1908) 181–83; "my special pets" in Ames 464.
13. Whitman (1899) 334.
14. "passenger pigeon's instinct": ibid. The two relevant facts were brought to my attention by David Blockstein.

15. Craig (1913) 95.
16. Pauly 162.
17. Deane (1908) 182.
18. Pauly 161.
19. Schorger (1955) 28; Pauly 162–64.
20. Ehrlinger 5; Cokinos 258.
21. Ehrlinger 5.
22. Ibid., 15; Cincinnati Zoo web page; Cokinos 256–59.
23. Schorger (1955) 29.
24. F. Thompson (1879) 265; Thompson (1881) 122.
25. Herman 78.
26. Ibid., 79.
27. Cokinos 259.
28. Deane (1909) 429.
29. Ibid.; Herman 79–80.
30. Cokinos 264.
31. Ibid.
32. Ibid., 266.
33. Ibid., 276–79.
34. Shufeldt 30.
35. Ibid., 31.
36. Ibid., 38.

CHAPTER 10: EXTINCTION AND BEYOND

1. Swarth 79; Rhoads 311.
2. French 33, 83.
3. Mershon Papers at Hoyt Library, University of Michigan, Ann Arbor.
4. Jason Weckstein, of the Field Museum and an expert on bird parasites, tells me that the possibility exists that some unknown disease may have affected the birds in ways that diminished their capacity to feed, breed, or conduct other vital funcitons without actually leaving telltale piles of corpses. At least one major study focusing on passenger pigeon DNA will be looking for diseases as well.
5. Bucher 7–9, 24, 19–20.
6. Josselyn in Wright (1910) 431; Kalm 58; Smith in Wright (1911) 428.
7. Buttons in Blockstein (2002) "Conservation and Management" 5. S. V. Wharram (68) tells of watching a small nesting colony in 1877 in Ohio, which he observed at leisure and where there is no mention of the birds' being disturbed (it seems to have been on his family's property). But apart from recalling only one egg per nest, he does not add much to the pool of facts, especially as to the success of the nesting effort. These smaller concentrations may have represented sink populations, birds drawn to an area seemingly suitable in habitat but possessing or missing some attribute that makes it difficult or impossible to reproduce. The most famous North American examples are neotropical migrants such as ovenbirds and wood thrushes that set up territories in Midwestern woodlands, but produce few if any offspring due to parasitism by brown-headed cowbirds.

8. Moose Factory in Mitchell 22; Schorger (1955) 36–43.
9. Schorger (1955) 212; Mitchell 139–40; Todd 270; Jackson and Jackson 769.
10. Mrazek, documentary treatment, http://e-int.com/messagefrommartha/.
11. Bucher 23; Blockstein (2002) "Conservation and Management" 4.
12. Halliday 159. That decline itself fostered increased mortality is related to the Allee effect, named for biologist Warder Clyde Allee. They relate to the "decline in individual fitness as low population size or density, that can result in critical population thresholds below which populations crash to extinction" (Courchamp et al., 2008 Oxford Scholarship Online). See also Reed 232–41.
13. Goodwin 176–78.
14. Cart 7–12. For a 1922 child's statement on bird conservation, see Greenberg (2008) 384.
15. Ibid., 66.
16. Blumm and Ritchie 126–27.
17. Hornaday 305–08.
18. Tober 159–62; *Missouri v. Holland.*
19. United States Fish and Wildlife Service.
20. Sixth mass extinction in Estes 301.
21. Blockstein (1989) 63–67; Myers 14–21.
22. Klehm 80–90; Lebbin et al. 318.
23. Lebbin et al. 328.
24. Phys.Org.
25. Lebbin 311; Loomis.
26. Arctic Climate Impact Assessment.
27. Greenberg (2002) 155–61, 174–76.
28. Lebbin et al. 307–9.
29. Ibid., 309–10.
30. Bat Conservation International; B. Miller A2.
31. United States Geological Survey.
32. Ibid., 296–98.
33. Many books, websites, and documentaries have been produced on what can be done to slow or reverse the negative impacts humans have on biodiversity. In addition, numerous private and public organizations actively address these and other environmental issues through research (both by professional and citizen scientists), education of adults and children, involvement in politics, and a broad array of outreach activities to engage as many people as possible. There is a role for everyone in this vital effort.

Bibliography

Allen, Elsa Guerdrum. "The History of American Ornithology Before Audubon." *Transactions of the American Philosophical Society*, n.s., 41 (pt. 3): 1951.

The American Field. "The Destruction of the Wild Pigeon." 17 (1882): 438.

———. "Final Effort to Find and Save from Extinction the Passenger Pigeon." February 5, 1910, 124–25.

American Society for the Prevention of Cruelty to Animals. http://www.aspca.org/about-us/history.aspx.

Ames, C. H. "Breeding of the Wild Pigeon." *Forest and Stream* 56 (June 15, 1901): 464.

Anbury, Thomas. *Travels Through the Interior Parts of America*. Vol. 1. London, 1791, 243–45.

Anderson, Rudolph. *The Birds of Iowa*. Davenport: Davenport Academy of Sciences, 1907.

Anonymous. "Finds Passenger Pigeon." June 1910. (Clipping with no other data sent by Karen Lund, of Genoa, Illinois, who found it in an old book.)

Anonymous. "The Past Participle in Pigeons." *Saturday Evening Post* 183 (October 15, 1910): 30.

Answers.com. "Railroads: Chronology." 1–27. http://www.answers.com/topic/railroads.

Antique. (Pigeon basket on cover.) August 1924. (In Jim Baillie Notes to Paul Hahn, Royal Ontario Museum, Toronto.)

Arctic Climate Impact Assessment. "Executive Summary" and "Key Finding #4: Animal species diversity, range and distribution will change." 2004. http://amap.no/acia/.

Armstrong, William. *Passenger Pigeons*. Pamphlet One. Blairstown, NJ, 1931.

Askins, Robert. *Restoring North America's Birds*. New Haven, CT: Yale University Press, 2000.

Atkinson, George. "A Review-History of the Passenger Pigeon in Manitoba." *Transactions of the Historical and Scientific Society of Manitoba* 68 (February 1905). Reprinted in Jim Blanchard, ed. *A Thousand Miles of Prairie: The Manitoba Historical Society and the History of Western Canada*. Winnipeg: University of Manitoba Press, 2002.

Audubon, John James. *Ornithological Biography, or An Account of the Habits of the Birds of the United States of America*. Vol. 1. Philadelphia, 1831, 319–26.

Audubon, Maria. *Audubon and His Journals*. Vol. 2. New York, 1897.

Bannon, Henry. *Stories Old and Often Told: Being Chronicles of Scioto County, Ohio*. Baltimore: Waverly Press, 1927.

Baraboo Republic (WI). "Pigeon Roost." June 21, 1882.

Barrows, Walter. *Michigan Bird Life*. Lansing: Michigan Agricultural College, 1912.

Bat Conservation International. "What We Do/White-Nose Syndrome." http://www.batcon.org/index.php/what-we-do/white-nose syndrome.html.

Beckner, Lucien. "The Last Wild Pigeon in Kentucky." *Transactions of Kentucky Academy of Science* 2 (1924–26): 55–56.

Belknap, Charles. *The Yesterdays of Grand Rapids*. Grand Rapids, MI: Dean Hicks Co., 1922.

Bent, Arthur Cleveland. *Life Histories of North American Birds of Prey*. United States National Museum Bulletin 167, vol. 2, 1937. (Reprinted by Dover Press in 1961.)

Benwell, J. *An Englishman's Travels in America*. London, 1853.

Bertram, Brian. "Living in Groups." In *Behavioral Ecology: An Evolutionary Approach*, edited by J. Krebs and N. Davies, 64–96. London: Blackwell, 1978.

Biggar, Henry, ed. *The Works of Samuel de Champlain*. Vol. 1. Toronto: Champlain Society, 1922–36. http://link.library.utoronto.ca/champlain/item_record.cfm?Idno=9_96821&lang=eng&query=The works of Samuel de Champlain, Vol. I&browsetype=Title&startrow=1.

Bishop, S. C. "A Note on the Food of the Passenger Pigeon." *Auk* 41 (1924): 54

Bishop, S. C., and A. H. Wright. "Note on the Passenger Pigeon." *Auk* 34 (1917): 208–09.

Blockstein, David. "A Federal Policy Is Needed to Conserve Biological Diversity." *Issues in Science and Technology* 5 (1989): 63–67.

———. "Lyme Disease and the Passenger Pigeon." *Science* 279 (1998): 1831.

———. "Passenger Pigeon." In *The Birds of North America Online*, edited by A. Poole. Ithaca, NY: Cornell Lab of Ornithology, 2002. doi:10.2173/bna.611.

Blockstein, David, and Harrison Tordoff. "Gone Forever: A Contemporary Look at the Extinction of the Passenger Pigeon." *American Birds* 39 (Winter 1985): 845–51.

Blumm, Michael, and Lucus Ritchie. "The Pioneer Spirit and the Public Trust: The American Rule of Capture and State Ownership of Wildlife." *Environmental Law* 50 (2005): 101–47.

BMR. "Treatment of Stool Pigeons." *American Field* 21 (April 26, 1884): 395–96.

Bogardus, Adam. *Field, Cover, and Trap Shooting*. New York: J. B. Ford and Co., 1879, 300–43.

Bond, Frank. "The Later Flights of the Passenger Pigeon." *Auk* 38 (1921): 523–27.

Bourne, Edward. *The History of Wells and Kennebunk*. Portland, Maine: 1875.

Brewster, William. "The Present Status of the Wild Pigeon (*Ectopistes mirgatorius*) as a Bird of the United States . . ." *Auk* 6 (1889): 285–91.

———. *The Birds of the Cambridge Region of Massachusetts*. Cambridge, 1906.

Brickell, John. *The Natural History of North Carolina*. Dublin, 1737.

Brinkley, Douglas. *The Wilderness Warrior: Theodore Roosevelt and the Crusade for America*. New York: Harper, 2009.

Brisbin, I. L. "The Passenger Pigeon: A Study in the Ecology of Extinction." *Modern Game Breeding* 4 (1968): 13–20.

Brown, William Griffee. *History of Nicholas County, West Virginia*. Markham, VA: Apple Manor Press, 2011. (First published 1954.)

Bruner, Lawrence. *Some Notes on Nebraska Birds*. Lincoln: State Journal Co., 1896.

Bucher, Enrique. "The Causes of Extinction of the Passenger Pigeon." *Current Ornithology* 9 (1992): 1–36.

Buckelew, Jay. E-mail to Bill Whan, February 13, 2012.

Buckingham, J. S. *The Slave States of America*. Vol. 2. London: Fisher, Son, and Co., 1842.

Bunnell, Lafayette. *Winona and Its Environs on the Mississippi in Ancient and Modern Days*. Winona, MN: Jones and Kroeger, 1897.

Butler, Amos. *The Birds of Indiana*. 22nd Annual Report of the Department of Geology and Natural Resources of Indiana for 1897. Indianapolis, 1898.

———. "Notes on Indiana Birds." *Proceedings of the Indiana Academy of Science*, 1899, 149–51.

———. "Some Rare Indiana Birds." *Proceedings of the Indiana Academy of Science*, 1902, 95–99.

———. "Some Notes on Indiana Birds." *Auk* 23 (1906): 271–72.

———. "Further Notes on Indiana Birds." *Proceedings of the Indiana Academy of Science*, 1912, 59–65.

Cart, Theodore. "The Struggle for Wildlife Protection in the United States, 1870–1900." Ph.D. diss., University of North Carolina (Chapel Hill), 1971.

Cartier, Jacques. *The Voyages of Jacques Cartier*. Edited and with an introduction by Ramsay Cook. Toronto: University of Toronto Press, 1993.

Cass County Republican (Dowagiac, MI). "Pigeons." May 31, 1860.

Casto, Stanley D. "Additional Records of the Passenger Pigeon in Texas." *Bulletin of the Texas Ornithological Society* 34 (2001): 5–16.

Catesby, Mark. *The Natural History of Carolina, Florida, and the Bahama Islands*. Vol. 1. London: Benjamin White, 1731, 23.

Chapman, Frank. "The Wild Pigeon at Englewood, New Jersey." *Auk* 13 (1896): 341.

Charles, Gordon. "A Big Black Cloud." *Traverse City Record Eagle* (MI), February 10, 1994.

Charlevoix Sentinel (MI), March 12, March 25, April 19, and April 26, 1878.

Chicago Field. "The Atoka Netting of Wild Pigeons." 15 (June 25, 1881): 314–15.

Chicago Tribune, August 17, 1872, 6; September 27, 1872, 6; January 2, 1875, 5; September 30, 1877, 7; and May 2, 1880. p. 3.

Cincinnati Zoo and Botanical Garden. http://cincinnatizoo.org/about-us/history-and-vision/.

Clark, Edward. "The Last of His Race?" *Chicago Tribune*, November 25, 1894, 44.

Clarke, James Freeman. *Memorial and Biographical Sketches*. Boston: Houghton Mifflin, 1878.

Clayton, Dale, and Roger Price. "Taxonomy of New World *Columbicola* (Phthiraptera: Philopteridae) from the Columbiformes (Aves), with Descriptions of Five New Species." *Annals of the Entomological Society of America* 92 (September 1999): 675–85.

Cleveland Plain Dealer. "Sky Rockets Among Pigeons." March 12, 1860.
Coale, Henry. "On the Nesting of *Ectopistes migratorius." Auk* 39 (1922): 254–55.
———. "Notes on *Ectopistes migratorius.*" *Auk* 39 (1922a): 255.
Cokinos, Christopher. *Hope Is the Thing with Feathers*. New York: Jeremy Tarcher/Putnam, 2000.
Cook, Ramsey, ed. *The Voyages of Jacques Cartier*. Toronto: University of Toronto Press, 1983.
Cook, William. "Large 19th Century Egg and Nest Collection Recently Discovered." *The Kingbird* 35 (Fall 1985): 247–50.
Cooper, James Fenimore. *The Pioneers*. New York: Signet, 1964. (Online version: http://xroads.virginia.edu/~UG02/COOPER/ch.22.html.)
Craig, Wallace. "The Expression of Emotions in the Pigeons. III. The Passenger Pigeon." *Auk* 28 (1911): 420–21.
———. "Recollection of the Passenger Pigeon in Captivity." *Bird-Lore* 15 (1913): 93–99.
Cruickshank, Helen, ed. *Thoreau on Birds*. New York: McGraw-Hill, 1964.
Cumberland Daily News (MD), October 9, 1872.
Cunningham, G. W. "Wild Pigeon Flights Then and Now." *Forest and Stream* 52 (March 25, 1899): 226.
Cutright, Paul. *Theodore Roosevelt: The Naturalist*. New York: Harper, 1956.
Czech, Kenneth. "Pottery: George Ligowsky and Modern Trapshooting." *Timeline* (Ohio Historical Society), March–April 1994, 22–27.
Daily Data (Green Bay, WI). "The Raid on the Pigeons." June 20, 1882.
Deane, Ruthven. "Some Notes on the Passenger Pigeon in Confinement." *Auk* 13 (1896): 234–37.
———. "Additional Records of the Passenger Pigeon (*Ectopistes migratorius*) in Wisconsin and Illinois." *Auk* 13 (1896a): 81.
———. "Additional records of the Passenger Pigeon." *Auk* 14 (1897): 316–17.
———. "The Passsenger Pigeon in Confinement." *Auk* 25 (1908): 181–83.
———. "The Passenger Pigeon—Only One Pair Left." *Auk* 26 (1909): 429.
———. "Abundance of the Passenger Pigeon in Pennsylvania in 1850." *Auk* 48 (1931): 264–65.
De L., H. W. "An Opinion on Trap-shooting." *Forest and Stream* 73 (April 22, 1880): 233.
Detroit Post and Tribune. "Birds of Passage: The Great Pigeon Nesting in Benzie County." April 29, 1880.
DeVoe, Thomas. *The Market Assistant*. New York: Hurd and Houghton, 1867.
Diamond, Seth, et al. "Hard Mast Production Before and After the Chestnut Blight." *Southern Journal of Applied Forestry* 24 (November 2000): 196–201.
Dictionary of Unitarian and Universalist Biography. "Henry Bergh." http://www25.uua.org/uuhs/duub/articles/henrybergh.html.
Dixon, E. C. "Last Great Flight of the Wild Pigeons in Wisconsin." *Platteville Journal*, April 16, 1930.
Dodge, E. S. "Notes from Six Nations on the Hunting and Trapping of Wild Turkeys and Passenger Pigeons." *Journal of Washington Academy of Sciences* 35 (1945): 342–43.

Dodson, Joseph. Clippings from Kankakee County Historical Society Archival Collection. Kankakee, IL.

Dunn, Rob. "On Parasites Lost." *Wild Earth*, Spring 2002.

Dury, Charles. "The Passenger Pigeon." *Journal Cincinnati Society of Natural History* 21 (1910): 52–56.

Eaton, Elon. *Birds of New York*. Pt. 1. Memoir 12. New York State Museum. Albany: University of State of New York, 1910.

Eenigenburg, Henry. *The Calumet Region and Its Early Settlers*. Chicago: Arrow Printers, 1935.

Ehrlich, Paul, and Anne Ehrlich. *Extinction: The Causes and Consequences of the Disappearance of Species*. New York: Random House, 1981.

Ehrlinger, David. *The Cincinnati Zoo and Botanical Garden: From Past to Present*. Cincinnati: Cincinnati Zoo and Botanical Gardens, 1993.

Ellsworth, Joshua, and Brenda McComb. "Potential Effects of Passenger Pigeon Flocks on the Structure and Composition of Presettlement Forests of Eastern North America." *Conservation Biology* 17 (December 2003): 1548–58.

Emmet County Democrat (Petoskey, MI). March 29, April 5, April 12, April 19, April 26, and May 3, 1878.

Estes, James et al. "Trophic Downgrading of Planet Earth." *Science* 333 (July 15, 2011): 301–306.

F. "A Trip to Buffalo in the St. Louis." *Western Literary Messenger* 9 (October 9, 1847): 149–50.

Faux, William. *Memorable Days in America: Being a Journal of a Tour to the United States*. London, 1823.

Fenton, W. N., and M. H. Deardorff. "The Last Passenger Pigeon Hunts of the Cornplanter Senecas." *Journal of Washington Academy of Sciences* 35 (1943): 289–315.

F. E. S. "Netting Wild Pigeons." *Forest and Stream* 43 (1894): 50.

Fleming, James. "Recent Records of the Wild Pigeon." *Auk* 20 (1903): 66.

———. "The Disappearance of the Passenger Pigeon." *Ottawa Field Naturalists* 20 (1907): 236–37.

Fond du Lac Commonwealth. "Among the Pigeons." May 20, 1871.

Fond du Lac Reporter, May 15, 1899.

Forbush, Edward Howe. "The Last Passenger Pigeon." *Bird-Lore* 15 (1913): 99–103.

———. *Game Birds, Wild-Fowl and Shore Birds*. Boston: State Board of Agriculture, 1916.

———. *Birds of Massachusetts and Other New England States*. Vol 2. Boston: Massachusetts Department of Agriculture, 1927.

Forest and Stream. "Passenger Pigeon." December 20, 1913, 792.

Fox, William. *The Bruce Beckons*. Toronto: University of Toronto Press, 1952.

Fradkin, Arlene. *Cherokee Folk Zoology*. New York: Garland Publishing, 1990.

Franklin, Wayne. *James Fenimore Cooper: The Early Years*. New Haven, CT: Yale University Press, 2007.

French, John C. *The Passenger Pigeon in Pennsylvania*. Altoona, PA: Altoona Tribune Company, 1919.

Friederici, Peter. "Passenger Pigeon Chewing Louse." *Wild Earth* 37 (Spring 1997): 37–38.

Fuller, Margaret. *Summer on the Lakes*. Boston: Little and Brown, 1844.

"G. D. Smith Succumbs." Obituary in unknown newspaper, October 26, 1940. In archives of Eastern Kentucky University, Richmond, KY.

Galveston Daily News (TX), November 17, 1875, 2; October 26, 1881, 3; and September 14, 1884, 2.

Garber, D. W. "The Passenger Pigeon in Ohio." *Ohio Conservation Bulletin* 20 (May 1956): 8, 28.

Gault, Benjamin. "The Passenger Pigeon in Aitkin County, Minnesota, with a Recent Record for Northeastern Illinois." *Auk* 12 (1895): 80.

GH. "Wild Pigeons." *Forest and Stream* 38 (1892): 79.

Giraud, J. P. *Birds of Long Island*. New York, 1844.

Gonterman, William. "The Passenger Pigeon in Western Kentucky." M. S. thesis, University of Kentucky, Lexington, 1929.

Goodwin, Derek. *Pigeons and Doves of the World*. Ithaca: Cornell University Press, 1983.

Grant, Anne. *Memoirs of an American Lady*. London, 1808.

Grant, Edward. "The Last Maryland Flight of the Passenger Pigeon." *Maryland Birdlife*, March–December 1951, 27–29.

Grant County Herald, May 1, 1847.

Greenberg, Joel. *A Natural History of the Chicago Region*. Chicago: University of Chicago Press, 2002.

———, ed. *Of Prairie, Woods, and Water*. Chicago: University of Chicago Press, 2008.

Griscom, Ludlow. "The Passing of the Passenger Pigeon." *American Scholar* 15 (1946): 212–16.

Guilday, John. "The Pigeons of Meadowcroft." In files of Meadowcroft Rockshelter and Historic Village, Avella, PA. Site managed by Heinz History Center, Pittsburgh, PA.

Guilday, John, and Paul Parmalee. "Vertebrate Faunal Remains from Meadowcroft Rockshelter, Washington County, Pennsylvania: Summary and Interpretation." In files of Meadowcroft Rock shelter and Historic Village, Avella, PA. Site managed by Heinz History Center, Pittsburgh, PA.

Hagar, Stansbury. "The Celestial Bear." *Journal of American Folklore* 13 (1900): 92–103.

Hall, James. *A Brief History of the Mississippi Territory*. Salisbury, NC, 1801.

Halliday, T. R. "The Extinction of the Passenger Pigeon *Ectopistes migratorius* and its Relevance to Contemporary Conservation." *Biological Conservation* 17 (1980): 157–62.

Hamel, Lynn. *A Taste of Old Madison*. Madison: Wisconsin Tales and Trails, 1974.

Harpel, Charles. Scrapbooks on "Chicago Men and Events," 1880s and 1890s—Serial A. In collection of Chicago Historical Society.

Harris, George. "The Life of Horatio Jones." *Publication of Buffalo Historical Society* (NY) 6 (1903).

Harris, Thaddeus Mason. *The Journal of a Tour into the Alleghany Mountains . . .* Boston, 1805.

Hartwick, L. M., and W. H. Tuller. *Oceana County: Pioneers and Business Men of Today*. Pentwater, MI: 1890.

Hedrick, U. P. *The Land of the Crooked Tree*. New York: Oxford University Press, 1948.

Hegde, S. N. "Composition of Pigeon Milk and Its Effect on Growth in Chicks." *Indian Journal of Experimental Biology* 11 (May 1973): 238–39.

Henninger, W. E. "A Preliminary List of the Birds of Middle Southern Ohio." *Wilson Bulletin* 9 (September 1902): 77–93.

Herman, William. "The Last Passenger Pigeon." *Auk* 65 (1948): 77–80.

Hewitt, John Hill. *Shadows on the Wall.* Baltimore: Turnbull, 1877.

Hine, Jane. "Game and Land Birds of an Indiana Farm." In *Biennial Report of the Commissioner of Fisheries and Game for Indiana*, 1911.

HM. *The American Field* 44 (December 7, 1895): 539.

Hodge, Clifton. "The Passenger Pigeon Investigation." *Auk* 28 (1911): 49–53.

———. "A Last Word on the Passenger Pigeon." *Auk* 29 (1912): 169–75.

Holland, Michigan, *Evening Sentinel.* "Hard Work is Formula for Young 82-Year-Old." September 14, 1944. (In archives of Lake Shore Museum Center, Muskegon, MI.)

Hollister, Ned. "Recent Record of the Passenger Pigeon in Southern Wisconsin." *Auk* 13 (1896): 341.

The Home Guide, or a Book by 500 Ladies. Richmond, IN: News Printing Co., 1881. (Compiled chiefly from "The Home" department of the *Chicago Daily Tribune.*)

Hornaday, William. *Our Vanishing Wildlife.* New York: Scribner's, 1913.

Hough, Emerson. "Wild Pigeons." *Forest and Stream* 39 (1892): 138.

———. "The Wild Pigeon." *Forest and Stream* 52 (1899): 88.

———. "A Genuine Wild Pigeon." *Forest and Stream* 53 (1899a): 248.

Howland, Catherine. *A Scrap Book of History of Decatur, Michigan, and Vicinity.* Vol. 1. Decatur, MI: Decatur Bicentennial Commission, 1976.

Hulbert, A. B., and W. N. Schwarze, eds. *David Zeisberger's History of the North American Indians.* Columbus: Ohio State Archaeological and Historical Society, 1910.

Hussey, Tacitus. "When Passenger Pigeons Were Plenty." *Forest and Stream* 82 (March 14, 1914): 5.

Indiana Farmer. "Pigeon Roosts Fifty Years Ago." 31 (February 22, 1896).

Indianapolis Star, March 17, 1934.

Jackson, H. Edwin. "Darkening the Sun in Their Flight: A Zooarchaeological Accounting of Passenger Pigeons in the Prehistoric Southeast." In *Engaged Anthropology*, edited by Michelle Hegmon and B. Sunday Eiselt. Ann Arbor: University of Michigan Museum, 2005.

Jackson, Jerome, and Bette Jackson. "Extinction: The Passenger Pigeon, Last Hopes, Letting Go." *Wilson Journal of Ornithology* 119 (2007): 767–72.

Johnson, W. S. "The Passenger Pigeon (*Ectopistes migratorius*) in Lewis County, N.Y." *Auk* 14 (1897): 88.

"John (X)." "Game Protection—Cruel Treatment of Stool Pigeons." *American Field* 21 (March 29, 1884): 299.

Jones, Clive, et al. "Chain Reactions Linking Acorns to Gypsy Moth Outbreaks and Lyme Disease Risk." *Science* 279 (February 13, 1998): 1023–32.

Jones, F. E. "Oldsters Meet for Birthdays." *Milwaukee Journal*, August 5, 1945.

Joyce, Clara. "The Ornithological Work of the Late Etta S. Wilson." *Indiana Audubon Society Yearbook* 9 (1936): 10–15.

Judd, Sylvestor. *History of Hadley.* Northampton, MA: Metcalf & Company, 1863.

Kalm, Pehr. "A Description of the Wild Pigeons Which Visit the Southern English Colonies in North America . . ." *Auk* 28 (1911): 53–66.

Kelly, Dave, et al. "An Intercontinental Comparison of the Dynamic Behavior of Mast Seeding Communities." *Population Ecology* 50 (2008): 339–42.

Kelly, Hugh. "The Great Pigeon Roost." *Baraboo Republic* (WI), May 3, 1871.

Kilbourn City Mirror. "The Great Pigeon Shoot." April 22, 1871.

———, May 13, 1871.

King, R. Ross. *The Sportsman and Naturalist in Canada*. London: Hurst and Blackett, 1866.

Kirtland, Jared. "New Habits of Birds." *Family Visitor*, July 8, 1851, 68.

Klehm, Daniel. "The Invisible Killer." *Bird Watcher's Digest* 14 (March–April 1992): 80–90.

Komar, Nicholas, and Andrew Spielman. "Emergence of Eastern Encephalitis in Massachusetts." *Annals of New York Academy of Sciences* 740 (December 15, 1994): 157–68.

Krech, Shepard. *Spirits of the Air: Birds and American Indians in the South*. Athens: University of Georgia Press, 2009.

Kriska, Nadine, and Daniel Young. "A Survey to Determine the Occurrence of the American Burying Beetle in Northern and Central Wisconsin." A joint proposal to the 1995–96 Wisconsin/Hilldale Undergraduate Faculty Research Fellowships, Madison, 1995.

Lake Shore Museum Center Archives. Unidentified news clipping showing wood duck. Lewis Cross file. Muskegon, MI.

Lalonde, R. G., and B. D. Roitberg. "On the Evolution of Masting Behavior in Trees: Predation or Weather?" *American Naturalist* 139 (1992): 1293–1304.

Larocque, A. "The Passenger Pigeon in Folklore." *Canadian Field-Naturalist* 44 (1930): 49–50.

Laudonniere, Rene. *Three Voyages* (translated, edited, and annotated by Charles E. Bennett). Gainesville: The University Presses of Florida. 1975.

Lawrence, Ed, and Peter Henkel. "The Michigan Pigeon Question: A Reply to Messrs. Martin and Turner." *Chicago Field* 11 (February 22, 1879): 25.

Laws of Michigan. 1875. No. 115, 149–50.

Lawson, John. *A New Voyage to Carolina*. Chapel Hill: University of North Carolina Press, 1967.

Lebbin, Daniel, et al. *The American Bird Conservancy Guide to Bird Conservation*. Chicago: University of Chicago Press, 2010.

Lee, Alfred. *History of the City of Columbus*. New York, 1892.

Leffingwell, William. *The Art of Wing Shooting*. Chicago and New York: Rand McNally, 1895.

Leonard, Henry. *Pigeon Cove and Vicinity*. Boston, 1873.

Lincecum, Gideon. "The Nesting of Wild Pigeons." *American Sportsman* 4 (June 27, 1874): 194–95.

Lincoln, C. C. "The Passenger Pigeon in Wisconsin." Mss. dated February 7, 1910, in Wisconsin Historical Society.

Littlefield, Daniel F., Jr. "'Roost Robbers' and 'Netters': Pigeoners in Indian Territory." *Chronicles of Oklahoma* 47 (Summer 1969): 154–59.

Litzke, Paul. "Wild Pigeons." *Forest and Stream* 54 (January 13, 1900): 24.

Loomis, Brandon. "Our Dying Forests: Dark Days for Whitebarks—and for Birds, Bears and Fish." *Salt Lake Tribune*. October 28, 2012. http://www.sltrib.com/sltrib/news/55109891-78/dyingforests-whitebark-pine-trees.html.csp?page=1.

MacKay, George. "Old Notes on the Passenger Pigeon (*Ectopistes migratorius*)." *Auk* 28 (1911): 261–62.

Mann, Charles C. *1491: New Revelations of the Americas Before Columbus*. New York: Vintage Books, 2005.

Mann, Charles L. "The Passenger Pigeon." *Annual Report of the Wisconsin Natural History Society, 1880–1881*, 1881, 45–47. (Translated from German by Carolyne Jaskula.)

Marsden, William. "What Has Become of the Wild Pigeon?" *Forest and Stream* 83 (1914): 146–47.

Martin, Edward. "Among the Pigeons." *Chicago Field* 10 (1879): 385–86.

———. "What Became of All the Pigeons." *Outing* 64 (July 1914): 478–81.

Massicotte, E.-Z. *Bulletin des Recherches Historiques* 34 (February 1928): 77–80. (Translated from French by Julia Innes.)

Mathew, Thomas. "The Beginning, Progress, and Conclusion of Bacon's Rebellion in Virginia in the Years 1675 and 1676." 1705. In Thomas Jefferson Papers, series 8, vol. 1:1–3, Library of Congress.

Maynard, Charles. *The Birds of Eastern North America*. Newtonville, MA, 1881.

McIlheny, E. A. "Major Changes in the Bird Life of Southern Louisiana During Sixty Years." *Auk* 60 (1943): 541–49.

McKenney, Thomas. *Sketches of a Tour of the Lakes*. Baltimore, 1827.

McKinley, Daniel. "A History of the Passenger Pigeon in Missouri." *Auk* (1977): 399–420.

McShea, William. "The Influence of Acorn Crops on Annual Variation in Rodent and Bird Populations." *Ecology* 8 (2000): 228–38.

Menard County Illinois History and Genealogy website. http://www.genealogytrails.com/ill/menard/1872/pg26.html.

Merritt, H. Clay. *The Shadow of a Gun*. Chicago: F. T. Peterson Co., 1904.

Mershon, William. *The Passenger Pigeon*. New York: Outing Publishing Co., 1907.

Michigan Tradesman, February 22, 1928.

Miller, Bob. [Decline of bats] *The News-Times* (Danbury, CT). December 18, 2010, A2.

Miller, W., and Ludlow Griscom. "Breeding of the Mourning Dove in Maine." *Auk* 37 (1920): 130.

Milwaukee Evening Wisconsin, June 20, 1882.

Milwaukee Journal, June 14, 1898; October 20, 1929; and September 18, 1935, 5.

Milwaukee Sentinel. "Millions of Pigeons. Ulster County, New York." May 7, 1872.

Missouri v. Holland. 252 U.S. 416 (1920).

Mitchell, Margaret. *The Passenger Pigeon in Ontario*. Toronto: University of Toronto Press, 1935.

Moody, Philip. "A Recent Record of the Wild Pigeon." *Bulletin of the Michigan Ornithological Club* 4 (September 1903): 81.

Mooney, James. "Myths of the Cherokees." *Nineteenth Annual Report of Bureau of American Ethnology, 1897–98*. Pt. 1. Washington, D.C., 1900.

Morse, Edward S. "Biographical Memoir of Charles Otis Whitman, 1842–1910." *National Academy of Sciences Biographical Memoirs* 7 (August 1912): 269–88.

Muskegon Chronicle (MI). "Paintings Preserve Colorful Lumber Era." August 3, 1937.

———. "Lewis Cross Dies at Spring Lake; Painted Pigeons, Many Scenics." April 6, 1951. (Both articles in archives of Lake Shore Museum Center, Muskegon, MI.)

Myers, Norman. "The Extinction Impending: Synergisms at Work." *Conservation Biology* 1 (May 1987): 14–21.

Nelson, Edward. "Birds of Northeastern Illinois." *Bulletin of the Essex Institute* 8 (1876): 120.

Nelson, Susan. "Prehistoric Passenger Pigeon Distribution Through Space and Time." B.A. thesis, Department of Anthropology and Sociology. University of Southern Mississippi, 2002.

Nessmuk [George Sears]. *Woodcraft*. 12th ed. New York, 1900.

Neumann, Thomas. "Human-Wildlife Competition and the Passenger Pigeon: Population Growth from System Destabilization." *Human Ecology* 13 (1985): 389–410.

Nevins, Allan, ed. *The Diary of Philip Hone*. New York: Dodd, Mead, 1927.

New York Times, June 21, 1881, 5, and June 22, 1881, 2.

Northern Tribune (Cheboygan, MI). March 9 and April 13, 1878.

Noss, Reed. *Forgotten Grasslands of the South*. Washington, D.C.: Island Press, 2012.

O'Callaghan, E. B. *The Documentary History of the State of New York*. Vol. 3. Albany, 1850.

Ohio Conservation Bulletin. "The Passing of the Passenger Pigeons." 19 (May 1955): 17, 28–29.

Ohio Historical Society Blog. "Fifth Most Embarrassing Moment in Ohio History." March 10, 2010. http://ohiohistory.wordpress.com/2010/03/10/now-for-number-5/.

Orlandini, John. "The Passenger Pigeon: A Seasonal Native American Food Source." *Pennsylvania Archaeologist* 66 (1996): 73–75.

Osprey (Osprey Company, Washington, D.C.) "Editorial Notes." 3 (1899): 12.

Ostfeld, Richard et al. "Of Mice and Mast: Ecological Connections in Eastern Deciduous Forests." *BioScience* 46 (May 1996): 323–329.

Palmer, Ralph. "Maine Birds." *Bulletin of the Museum of Comparative Zoology* (Harvard College) 12 (July 1949): 298–99.

Parmalee, Paul. "Remains of Rare and Extinct Birds from Illinois Indian Sites." *Auk* 75 (1958): 169–76.

———. "Animal Remains from the Modoc Rock Shelter Site, Randolph County, Illinois." In *Summary Report of Modoc Rock Shelter, 1952, 1953, 1955, 1956*. Illinois State Museum Reports of Investigations no. 8. Springfield, 1959.

Pauly, Philip. *Biologists and the Promise of American Life*. Princeton, NJ: Princeton University Press, 2000.

Paxson, Henry. "The Last of the Wild Pigeon in Bucks County." *Bucks County Historical Society Collection* 4 (1917): 367–82.

Peterson, Eugene. "The History of Wildlife Conservation in Michigan." Ph.D. diss. University of Michigan, 1952.

Phys.Org. "Overfishing Pushes Tuna Stocks to the Brink." September 8, 2012. http://phys.org/news/2012-09-overfishing-tuna-stocks-brink-experts.html

Potter Journal (Coudersport, PA), April 15, May 13, and June 10, 1880. Clippings in the Coudersport, PA, Historical Society.

Powell, J. H. *Bring Out Your Dead: The Great Plague of Yellow Fever in Philadelphia in 1793*. Philadelphia: University of Pennsylvania Press, 1993.

Powers, Grant. *Historical Sketches of the Discovery, Settlement, and Progress of Events in the Coos Country and Vicinity*. Haverhill, NH, 1880.

Prairie Farmer (Chicago). "Viciousness of Pigeons." 12 (February 1852): 83.

Price, Jennifer. *Flight Maps*. New York: Basic Books, 2000.

Purdue, James. "The Father and Daughter of the Oliver S. Biggs Museum of Natural History." *Living Museum* 51 (1989): 51–54.

R. "The Pigeon Trade." *Plattsburgh Republican* (NY), August 2, 1851.

Rader, Walter. *Indianapolis Star*. March 17, 1934. Reprinted in Wilson, "Historical Notes on DuBois County," vol. 16: 442.

Radin, Paul. "The Winnebago Tribe." *Thirty-Seventh Annual Report of the Bureau of American Ethnology*. Washington, D.C., 1923.

Raithel, Christopher. *American Burying Beetle (*Nicrophorus americanus*): Recovery Plan*. Newton Corner, MA: U.S. Fish and Wildlife Service, 1991.

Randolph, Vance. *We Always Lie to Strangers: Tall Tales from the Ozarks*. New York: Columbia University Press, 1951.

Raper, Frank. "Pigeons." *Columbus Dispatch*, April 9, 1939.

Ray, Cap. "Early Days in Backwoods." *Lakeland Times* (Minocqua, WI), March 9, 1950.

Reed, J. Michael. "The Role of Behavior in Recent Avian Extinctions and Endangerments." *Conservation Biology* 13 (April 1999): 232–41.

Reeve, Simon. "Going Down in History." *Royal Geographic Society Magazine* 73 (March 2001): 60–64.

Revoil, Benedict Henry. *The Pigeons: A Story and a Prophesy*. Translated by William Benignus. Altoona: Pennsylvania Alpine Club, 1926.

Rhoads, Samuel. "The Wild Pigeon . . . on the Pacific Coast." *Auk* 8 (1891): 310–12.

Roberts, Thomas. *The Birds of Minnesota*. Vol. 1. Minneapolis: University of Minnesota Press, 1932.

Robinson, E. Arthur. "Conservation in Cooper's *The Pioneers*." *PMLA* 82 (1967): 564–78.

Roney, Henry. "Among the Pigeons: A Description of the Pigeon Nesting of 1878 and the Work of Protection Undertaken by the East Saginaw and Bay City Game Protection Clubs." *Chicago Field* 10 (1879): 345–49.

Rumer, Tom. *Unearthing the Land: The Story of Ohio's Sciota Marsh*. Akron: University of Akron Press, 1999.

Rupp, William. "Bird Names and Bird Lore Among the Pennsylvania Germans." *Pennsylvania German Society Proceedings and Addresses* (Norristown, PA) 52 (1946).

Sage, John, et al. *The Birds of Connecticut*. Hartford: State Geological and Natural History Survey of Connecticut, 1913.

Schaff, Morris. *Etna and Kirkersville*. Boston: Houghton Mifflin, 1905.

Schapper, Ferdinand. "Southern Cook County and History of Blue Island Before the Civil War." Vol. 1. Typed manuscript, 1917. Chicago Historical Society.

Scherer, Lloyd, Jr. "The Passenger Pigeon in Northwestern Pennsylvania." *Cardinal* 5 (1939): 25–42.

Schoolcraft, Henry Rowe. *Narrative Journal of Travels Through the Northwestern Regions of the United State Performed as a Member of the Expedition Under Governor Cass in the Year 1820*. Albany, 1821.

Schorger, A. W. "The Great Wisconsin Passenger Pigeon Nesting of 1871." *Proceedings of the Linnaean Society of New York* 48 (1937): 1–26.

———. "Unpublished Manuscripts by Cotton Mather on the Passenger Pigeon." *Auk* 55 (1938): 471–77.

———. *The Passenger Pigeon: Its Natural History and Extinction*. Madison: University of Wisconsin Press, 1955.

Scott, Walter, ed. *Silent Wings: A Memorial to the Passenger Pigeon*. Madison: Wisconsin Society for Ornithology, 1947.

Seton, Ernest Thompson. "The Birds of Manitoba." *Proceedings United States National Museum* 18 (1891): 522–23.

Sharkey, Reginald. *The Blue Meteor*. Petoskey, MI: Little Traverse Historical Society, 1997.

Shufeldt, R. W. "Anatomical and Other Notes on the Passenger Pigeon . . . Lately Living in the Cincinnati Zoological Gardens." *Auk* 32 (1915): 29–41.

Sibley, John L. *A History of the Town of Union, in the County of Lincoln, Maine*. Boston, 1851.

Simmons, G. F. *Birds of the Austin Region*. Austin: University of Texas, 1925.

Smith, George D. "The Tragedy of the Passenger Pigeon." Unpublished and undated memoir, 1–6. Archives of Eastern Kentucky University, Richmond.

Smith, Katherine, et al. "Evidence for the Role of Infectious Disease in Species Extinction and Endangerment." *Conservation Biology* 20 (2006): 1349–57.

"Snap Shot." "Peace and Pigeons in Wisconsin." *Wilkes' Spirit of the Times* 12 (May 27, 1865): 194.

Snyder, Dorothy. "The Passenger Pigeon in New England." *Old Time New England*, no. 3 (1955): 3–14.

Sork, Victoria, et al. "Ecology of Mast-Fruiting in Three Species of North American Deciduous Oaks." *Ecology* 74 (1993): 528–41.

South Shore Country Club Magazine: Golden Anniversary. "A Day's Hunting in 1871 on the Club's Ground." 42 (August 1956): 34.

Stanstead. "Spring Notes." *Forest and Stream* 40 (1893): 403.

Steele, Zulma. *Angel in Top Hat*. New York: Harper and Brothers, 1942.

St. Joseph Traveler (MI). "Game Hunt." October 26, 1878.

Stone, Fanny. *Racine and Racine County, Wisconsin*. Vol. 1. Chicago: S. J. Clark, 1916.

Stone, Witmer. *Birds of New Jersey*. Trenton: J. L. Murphy, 1909.

Stratton-Porter, Gene. "The Last Passenger Pigeon." In *American Earth*, edited by Bill McKibben. New York: Library of America, 2008.

Swanson, Evadene. "The Use and Conservation of Minnesota Game, 1850–1900." Ph.D. diss. University of Minnesota, 1940.

Swarth, Harry S. [HSS]. "Publications Reviewed: Notes on the Passenger Pigeon." *Condor* 13 (March–April 1911): 79.

Thomas, Edward. "Trap Shooting in the Old Days." *Outing* 66 (1915): 368–72.

Thomas, Lately. *Delmonico's: A Century of Splendor*. Boston: Houghton Mifflin, 1967.

Thompson, Frank. "Incubation Under Difficulties." *Forest and Stream* 12 (May 8, 1879): 265.

———. "Breeding of the Wild Pigeon in Confinement." *Nuttall Bulletin* 6 (1881): 122.

Thompson, W. W. *The Passenger Pigeon*. Coudersport, PA, 1921.

Tober, James. *Who Owns the Wildlife?* Westport, CT: Greenwood Press, 1981.
Todd, W. E. C. *Birds of Western Pennsylvania*. Pittsburgh, 1940.
"Tom Tramp." "A Pigeon Roost." *Rod and Gun* 8 (June 3, 1876): 149.
Townsend, Charles Wendell. "Passenger Pigeon." In Arthur Cleveland Bent's *Life Histories of North American Gallinaceous Birds*. Bulletin of the U.S. National Museum no. 162. Washington, D.C., 1932, 379–402.
Trautman, Milton. *The Birds of Buckeye Lake, Ohio*. Ann Arbor: University of Michigan Press, 1940.
Traverse, Robert, ed. "The Passenger Pigeon Becomes Extinct." *A Scrapbook History of Early Decatur, Michigan*, 1976, 1411.
Turner, A. B. "The Michigan Pigeon Question: Mr. Turner's Reply to Professor Roney." *Chicago Field* 10 (February 1, 1879): 401–02.
Twain, Mark. *Autobiography*. Vol. 1. New York: Harper & Brothers, 1924.
United States Fish and Wildlife Service Species Accounts Online. http://www.fws.gov/species/species_accounts/bio_swan.html.
United States Geological Survey, Fort Collins Science Center. "White-Nose Syndrome Threatens the Survival of Hibernating Bats in North America." http://www.fort.usgs.gov/WNS/.
Upton, William Treat. *Anthony Philip Heinrich*. New York: AMS Press, 1967.
Viroqua Censor (Wis). "The Pigeon Roost." December 1, 1880.
Ward, Marion. "Four Cross Brothers Are Well Known in Michigan." *Grand Rapids Herald*, May 11, 1940. (Lewis Cross file. Archives of Lake Shore Museum Center, Muskegon, MI.)
Watson, John. *Annals of Philadelphia and Pennsylvania*. Vol. 2. Philadelphia, 1857.
Webb, Sara. "Potential Role of Passenger Pigeons and Other Vertebrates in the Rapid Holocene Migrations of Nut Trees." *Quaternary Research* 26 (1986): 367–75.
Webber, C. W. "The Wild Pigeon." *Arthur's Home Magazine*, April 1854, 305–08.
Welsh, William. "Passenger Pigeons." *Canadian Field-Naturalist* 39 (1925): 165–66.
Wharram, S. V. "The Passenger Pigeon in Ohio." *Bird-Life* 39 (August 30, 1943): 65–68.
Wharton, Richard. "The Extinction of the Passenger Pigeon—an American Tragedy." Unpublished and undated memoir, a copy of which was conveyed by Wharton (Joaquin, TX) to David Wolf (Nacogdoches, TX), who in December 2001 gave a copy to Stan Casto (Seguin, TX), who gave a copy to me.
Wheaton, J. M. "Report on the Birds of Ohio." *Report of the Geology of Ohio* 4 (1882): 442.
Whitewater Register (WI), June 3, 1880.
Whitman, Charles Otis. "Animal Behavior." *Biological Lectures from the Marine Biological Laboratory, Wood's Hole, Massachusetts*. Boston, 1899.
———. *Posthumous Works of Charles Otis Whitman*. Washington, D.C.: Carnegie Foundation, 1919.
Whittaker, David. "The Days of the Wild Pigeon." *Outdoor America*. July 1928, 28–30.
Wicks, J. B. *My Bird Parishioners*. Paris Hill, NY: 1897.
Widmann, Otto. *A Preliminary Catalog of the Birds of Missouri*. St. Louis: Academy of Science of St. Louis, 1907.
Wilcove, David. "In Memory of Martha and Her Kind." *Audubon* 91 (1989): 52–55.

Williams, Charles. "Why Are There So Few Insect Predators of Nuts of the American Beech (*Fagus grandifolia*)?" *Great Lakes Entomologist* 40 (2007): 140–53.

Wilson, Alexander. *American Ornithology; or, The Natural History of the Birds of the United States*. Boston, 1839.

Wilson, Etta. "Personal Recollections of the Passenger Pigeon." *Auk* 51 (1934): 157–68.

———. "Additional Notes on the Passenger Pigeon." *Auk* 52 (1935): 412–13.

———. "Kin-ne-quay." Undated source document, 1–4. the Holland Museum Archives and Research Library, Holland, MI.

Wilson, George. "Historical Notes on DuBois County." Vols. 14, 16. Jasper, IN: Dubois County Historical Society.

Winter, William. *Shadows of the Stage*. 2nd ser. 1894. http://www.wayneturney.20m.com/boothjb.htm.

Wood, J. Claire. "The Last Passenger Pigeons in Wayne County, Michigan." *Auk* 27 (1910): 208.

Wood, Norman. *The Birds of Michigan*. No. 75. Miscellaneous Publications of the Museum of Zoology of the University of Michigan, Ann Arbor, 1951.

Woodruff, Frank. *The Birds of the Chicago Area*. Bulletin no. 6 of the Natural History Survey. Chicago Academy of Sciences, 1907.

Wright, Albert Hazen. "Some Early Records of the Passenger Pigeon." *Auk* 27 (October 1910): 428–43.

———. "Other Early Records of the Passenger Pigeon." *Auk* 28 (July 1911): 346–66.

———. "Other Early Records of the Passenger Pigeon." *Auk* 28 (October 1911): 427–49.

Young, Duane. "Ecological Considerations in the Extinction of the Passenger Pigeon . . . Heath Hen . . . and the Eskimo Curlew." Ph.D. diss. University of Michigan, 1953.

Index

Note: Italic page numbers refer to illustrations.

A Note on the Author

Joel Greenberg is a research associate of both the Chicago Academy of Sciences Peggy Notebaert Nature Museum and the Field Museum. He has taught courses on natural history for the Morton Arboretum, Brookfield Zoo, and Chicago Botanical Garden, and is the author of *Of Prairie, Woods, and Waters: Two Centuries of Chicago Nature Writing*; *A Natural History of the Chicago Region*; and *A Birder's Guide to the Chicago Region* (with Lynne Carpenter). His articles or reviews have been published in *Science*, *Birder's World*, and *Environmental History*, and he is a featured blogger on the website Birdzilla.com. He cohosted a radio show on the outdoors on WKCC. A retired member of the Illinois Bar, Greenberg has received the Protector of the Environment Award (1997, Chicago Audubon Society) and Environmental Leadership Award (2004, Institute for Environmental Science and Policy, University of Illinois at Chicago).

Beyond writing *A Feathered River Across the Sky*, Greenberg is working with David Mrazek on creating a documentary on passenger pigeons and extinction called *From Billions to None: The Passenger Pigeon's Flight to Extinction*. These projects are all major elements of Project Passenger Pigeon, an international effort which Greenberg helped bring about.

An avid birder since the age of twelve, Greenberg, his wife, and their stuffed passenger pigeon, Heinrich, reside in the Chicago area.